Future of Business and Finance

The Future of Business and Finance book series features professional works aimed at defining, analyzing, and charting the future trends in these fields. The focus is mainly on strategic directions, technological advances, challenges and solutions which may affect the way we do business tomorrow, including the future of sustainability and governance practices. Mainly written by practitioners, consultants and academic thinkers, the books are intended to spark and inform further discussions and developments.

Anna K. Zarkada · Yioula Melanthiou

Future Management for Sustainable Prosperity

Regenerating Mindsets, Rethinking Strategic Agility

Springer

Anna K. Zarkada
Department of Communication
and Marketing
Cyprus University of Technology
Limassol, Cyprus

Yioula Melanthiou
Department of Communication
and Marketing
Cyprus University of Technology
Limassol, Cyprus

ISSN 2662-2467 ISSN 2662-2475 (electronic)
Future of Business and Finance
ISBN 978-3-032-15867-3 ISBN 978-3-032-15868-0 (eBook)
https://doi.org/10.1007/978-3-032-15868-0

This Springer imprint is published by the registered company Springer Nature Switzerland AG
The registered company address is: Gewerbestrasse 11, 6330 Cham, Switzerland

Preface—The Next Normal, But Different

One of the traditional aspects of organisations has always been dealing with uncertainty. The twenty-first century is unique in its sustained presence of this uncertainty. Once thought of as being a temporary disturbance on the road to stability, the term VUCA (an environment characterised by Volatility, Uncertainty, Complexity, and Ambiguity) has become more of a chronic disorder: Permacrisis. Overlapping economic downturns, ecological crises, technological revolutions, pandemics, and geopolitical crises have long set in. They overlap with one another, cascade down, and multiply with each other. For managers, volatility is no longer a temporary bump on the road to stability; it has become the norm.

Given this, the core question of this book is, "How can an organisation thrive when stability is the exception and not the rule?" The answer we propose is "Future Management"; a whole new mindset revolving around three mutually connected pillars: sustainable prosperity, values-driven value creation, and strategic agility. Together, these three provide managers with a road map on how to navigate the pressures of profitability and all the stewardship obligations towards people, societies, and the planet.

Many books offer an outdated set of best practices. This book strives for a full-blown rethinking of management: what management is supposed to be, the assumptions made about it, and its position in society. One of the most pressing issues of our time is to examine management beyond the limelight of technical discipline of resource allocation and efficiency. Management at its root should be a cultural and ethical practice, a resurrected mindset, a higher understanding of interconnected systems, and the courage to design organisations committed to delivering prosperity for all.

The term VUCA was coined in an attempt to explain the new-age challenges for management. Not even the usage of the term could be enough to suggest that classic managerial tools were not very useful, but VUCA implied that there would be some turbulence which would be episodic, and a period of calm would set in thereafter. Crises in the present world are no longer discrete events, and while our world shall still remain VUCA, what has changed is that challenges and changes are now permanent, intertwined, and systemic.

The permacrisis setting provided here for the rest of the book highlights that problems and limitations arise from single-mindedly pursuing narrowly viewed competitiveness or growth-at-any-cost business strategies. This brings in the need to expand the scope of management: from being concerned with short term quarterly earnings to long term sustainability, from focusing on shareholder primacy toward multi-stakeholder responsibility, and from rigid control toward agile and flexible approaches.

The Book's Core Argument

The proposition of this book is simple yet demanding: the solution to the problems facing management today, and those emerging in the future, requires a critical turnaround. Future Management therefore has to be embraced as a paradigm. This paradigm rests on three pillars:

1. Sustainable Prosperity: development that advances the well-being of people, organisations, and societies without draining the environment nor sacrificing the needs of future generations.
2. Values-Driven Value Creation: employing technology, innovation, and organisational purpose through systematic action to co-create value that is fair, inclusive, and devoid of detrimental externalities.
3. Strategic Agility: the embedding of adaptability, foresight, and resilience in the very core of organisational design and decision-making.

Future managers will need to possess a strong will to pursue shared prosperity instead of reckless growth-chasing.

Structure of the Book

This book is divided into three parts across four chapters. The first introduces the challenges a manager faces in an ever-ongoing crisis; the second establishes the core components of Future Management, whereas the third deals with the repercussions on leadership, organisations, and society.

Chapter 1: sets the conceptual foundation and diagnoses the present setting. It explains all the issues caused by political unrest, weak economies, social change, rapid technological advances, change in the law, and the environment that together create a never-ending change. It then finally draws attention to the almost obsolete competitive and growth models, demonstrating that management needs to address structural problems in a permacrisis situation.

Chapter 2 features pillar number one: sustainable prosperity. In this chapter, prosperity is therefore redefined, no longer as an accumulation of wealth, but rather as some sort of balance, interdependence, and sustainability among generations.

The chapter integrates economic debates and philosophical traditions to suggest that managers must bear responsibility for sustainability and universal prosperity.

Chapter 3 addresses the second pillar of future management: value and values. The discussion shifts to the organisation's operations, focusing on value for investors, employees, customers, and society in general. Values should guide the organisation toward its legitimate pursuit of fairness and long-term trust. The argument is that without values, value becomes extractive and destructive, but with values embedded throughout operations, organisations become useful as well as profitable and sustainable.

Chapter 4 presents the third pillar of future management: strategic agility and capability building. Without any flexibility once in a permacrisis, the end is certain. The chapter presents companies which were able to rebound from certain crises through agility as a core capability and, contrary to the more static models, demonstrated adaptability, foresight, and resilience embedded in their culture, processes, and strategy.

These four chapters join one narrative: managers can no longer work on assumptions and methods that have been passed on to them from their predecessors. To be effective and relevant for the twenty-first century, managers need to embrace the Future Management paradigm and work towards Sustainable Prosperity.

What This Book Offers

This book offers critical perspectives and integrative frameworks. Drawing from management theory, economics, philosophy, and practice, we have aimed to provide readers with a narrative that acknowledges today's challenges but also offers a positive reflection; a shift in mindset from what managers do, to why managers do what they do. This message is reinforced and made practical through the "Future Management in Practice" reflections spread throughout the book, which make theory applicable to day-to-day work, thus easing implementation.

For decades, business schools have trained managers to maximise shareholder value, optimise efficiencies, and compete aggressively. These tactics are certainly still important but are no longer sufficient for navigating through a permacrisis. Future managers must be independently capable of engaging with societal debates, anticipating systemic risks, and leading with values rather than just analytics.

The following chapters do not pretend to offer all of the required solutions ready on a plate. They provide frameworks, ideas, and questions to help Future Managers rethink their role. What unites these chapters is their assumption that management can no longer exist as a limited, technocratic discipline concerned with efficiency. Rather, management needs to become a stewardship of people, organisations, and ecosystems; stewardship based on values working toward sustainable prosperity and agile enough to constantly respond to disruptions. If management continues to cling to outdated concepts such as infinite growth and short-term profit, they will worsen the very problems they are trying to solve. But if management embraces

Future Management ideas, this would be the force of regeneration and resilience helping societies and organisations to sail smoothly through the age of permacrisis.

Limassol, Cyprus

Anna K. Zarkada
Yioula Melanthiou

Acknowledgements The authors gratefully acknowledge the contribution of Professor Luiz Moutinho in the development of the Future Management concept and his instrumental encouragement for writing this book. The quality and relevance of the text have been improved by comments on early drafts of Chapters 1–3 by Vasiliki Zarkada, Isadora Fraser, and Iannis Zarkadas. Last, but not least, our heartfelt appreciation goes to our Editor Jianlin Yan who skilfully and sensitively guided us through rough patches.

Competing Interests The authors have no competing interests to declare that are relevant to the content of this manuscript.

Contents

About the Author

Dr. Anna K. Zarkada is an Associate Professor of Integrated Marketing Communications and Head of the Department of Communication and Marketing, Cyprus University of Technology. She holds a B.Sc. (1st Class Honours) in Business Administration from the Athens University of Economics and Business, Greece, an M.Sc. by Research in International Marketing from the University of Manchester, UK, and a Ph.D. in Service Science and Ethics from the Queensland University of Technology, Australia.

She has taught in leading Greek, UK, USA, Australian, Indian, Russian, Cypriot, and Japanese Higher Education Institutions. She is the Editor of the books *Marketing Solutions to the Challenges of a VUCA Environment* and *Empowering Digital Transformation* which were published by Springer and the Author of *Integrated Marketing Communications for Sustainable Prosperity: Theory and Practice* which is in print with Routledge. Her research on digital marketing, business ethics and philosophy, services, and cross-cultural communication has been published in over 130 articles and conference papers in top-tier journals and international conferences. She has served as conference chair and member of the editorial board, reviewer, and guest editor of numerous prestigious journals.

She has received several best paper, article, and book chapter awards as well as a reviewer of the year award. She is an active consultant and trainer for leading global companies, small businesses, and governmental and non-profit organisations. She is also involved in the promotion of Music and Arts Education.

Anna has written Chapters 1–3.

Dr. Yioula Melanthiou is an Assistant Professor of Marketing at the Cyprus University of Technology. Her research interests focus on Augmented Reality (AR), Social Media Marketing, and Consumer Behaviour, fields in which she has numerous publications and presentations. Dr. Melanthiou brings over two decades of academic experience, having held leadership roles including Head of the Marketing Department and Director of Doctoral Studies. Prior to her academic career, she served as Research Director and Marketing Consultant for a multinational company. She is deeply committed to teaching and mentoring undergraduate and doctoral students and has actively participated in many European funded research programs.

Yioula has written Chapter 4 and has tied the whole book together with the Preface and the Future Management in Practice boxes which run throughout the book.

From a VUCA Environment to a Permacrisis World

1

Acronyms

PESTLE	Political, Economic, Social, Technological, Legal, Environmental
VUCA	Volatile; Uncertain, Complex, Ambiguous
3BL	Triple Bottom Line

When the idea for this book was being developed, we came up with a list of challenges management faces in the third decade of the twenty-first century. We planned to discuss the major Political, Economic, Social, Technological, Legal, and Environmental (PESTLE) factors and how these create a Volatile, Uncertain, Complex and Ambiguous (VUCA) global environment for Multinational Corporations (MNCs) and large internationalised corporations, but also for localised organisations, and even the smallest businesses. As the last of the COVID-19 restrictions were being lifted, it was clear that the 'new normal' was manifesting itself in unprecedented, not so rosy, ways. In 2022 the "feeling of living through a period of war, inflation, and political instability" was so pervasive to be given a name and a place in the dictionaries. If you feel that you are in an "ongoing state of uncertainty and worry" it's not you; it is the world. We are in "[a]n extended period of instability and insecurity, especially one resulting from a series of catastrophic events": permacrisis (Bushby, 2022).

Catastrophic events continued to pile up through 2023 and made it clear that the permacrisis was not just a random agglomeration of unfortunate occurrences. Permacrisis manifested as a complete system failure caused by the way growth, policy management, and governance had been approached at both domestic and global level. Widespread economic instability, recurring climate emergencies, rising inequality, and dangerous geopolitical tensions were now recognised as just the symptoms of a broken system producing a worrisome global outlook (Brown et al., 2023).

A. K. Zarkada and Y. Melanthiou, *Future Management for Sustainable Prosperity*, Future of Business and Finance, https://doi.org/10.1007/978-3-032-15868-0_1

The grim outlook persisted throughout 2024. The 'new normal' looked like a flock of Black Swans thick enough to cast an opaque shadow over peace, security, and prosperity. At any time during the year, we were experiencing the concurrence of, far too many to even list, catastrophic events. Each one of these events was characterised by "rarity, extreme impact, and retrospective … predictability" (Taleb, 2010, p. 21). All of them together made the permacrisis look as solid as a rock and 2024 "one of the worst years in UNICEF's history" (English, 2024).

Nothing improved during 2025. It uncertainly and ambiguously progressed from one illogical and utterly bloody conflict to the next. According to a "global CEO advisory firm" (Kajiwara & Piccoli, 2024):

> One thing is clear: beyond individual events, the geopolitical fault lines continue to shift, with significant consequences for international trade and investment. The global economic environment is pivoting toward greater securitization, regional fragmentation and a retreat from the rule-based frameworks that defined much of the post-war political and economic order. International frictions are set to persist at levels unseen since the Cold War's conclusion, reshaping trade flows, investment strategies and the more general risk landscape. In this era of heightened uncertainty, the map of opportunities and challenges for firms is being fundamentally redrawn. (Kajiwara & Piccoli, 2024)

There is no reason to believe that 2026 will be any less of a dystopian version of 'normal' and as 'new' as 1913—if you focus on technological disruption or nationalist rhetoric—or 1939—in terms of geopolitics and the rise of authoritarianism. It is obvious that the permacrisis, and also genocides, wars, and the threat of a nuclear apocalypse are here to stay, and we can only hope this, whatever *this* might be, *is* the worst of it.

Pandora's box (Fig. 1.1) has been opened, the evils have been released, and the hope trapped inside might make individual woes bearable but does not solve the problems or reverse the path to an unsustainable future. Even though it would take a psychic or a prophet—and certainly not an academic—to forecast the next day, let alone the outcome of current events within the strategic planning horizon, we believe that by understanding the causes of events and regenerating mindsets to address them we can prevent the next catastrophic development, or at least manage its repercussions, and thus we can help managers guide organisations through the permacrisis.

There are, of course, no Messiahs and things will not magically improve overnight. Neither pretending that the 'new normal' *is* normal, or desirable, will make it so. We need to understand what breeds the Black Swans and gives a crisis—an event which should, by definition, be an unexpected and transient one—its permanence. In this chapter we discuss the key problems contemporary management faces, attempt to trace their roots, and examine their immediate consequences. Our starting point is our concurrence with earlier analyses (Etzioni, 1988; Panigyrakis & Zarkada, 2014a; Schumacher, 1973; Sen, 1988) which place the "consciously and purposefully amoralistic neoclassical paradigm" (Panigyrakis & Zarkada, 2014a) at the root of the current Volatile Uncertain Complex

Fig. 1.1 Pandora, oil on canvas by Odilon Redon, 1910/1912, from the Chester Dale Collection, The National Gallery of Art, Washington DC. Available from https://www.nga.gov/collection/art-object-page.46531.html through an Open Access licence

and Ambiguous (VUCA) Political Economic Social Technological Legal and Economic (PESTLE) conditions. We argue that it is pervasive amorality—the broken link between moral judgement and the motivation to comply with it which cannot "be explained away as … irrationality, conceptual incompetence, or insincerity" (Rosati, 2016)—which provides the breeding ground of a pervasive but faulty mindset of prioritising profits over people and the planet. This amoralistic stance towards business activity guides a few major players' persistently irresponsible actions and lulls the passive majorities into accepting the consequences of business and political leaders' self-serving decisions as their fate.

The common characteristic of all organisations which distinguishes them from all other forms of human groupings is structure: people's skills, thoughts, and actions are structured to achieve a common goal within given PESTLE structures. In business organisations, managers are trained to plan within given legal and regulatory frameworks, to efficiently organise and mobilise resources within their control, and to take calculated risks based on solid market intelligence. Even to be able to think outside the box, one needs to be aware of the structure of the box. The twenty-first-century permacrisis box is characterised by high political risk and low information integrity. Its loose and porous structures are unable to contain what escaped Pandora's box and support what is still left at the bottom of it. Thus, in their daily practice, managers are most likely to face circumstances they are not adequately prepared for.

We argue that people are not helpless. We can change the structures and fix the broken system. First we need to understand what we did wrong and why. Then we need to shift our perspective to mend our ways. In the following sections we pinpoint the mentalities that need changed for the way out of the permacrisis to become visible and explore how understanding events and their causes sets the basis for the regeneration of mindsets. If we understand the faults in our thinking which caused the problems we are facing, we can facilitate the regeneration of mindsets, embrace technology, and develop the strategic agility required for reversing the damages and paving the way forward in a manner conducive to sustainable prosperity for all.

1.1 The Political Environment: Upheavals and Their Effects on the Economy, Society, and Business Operations

Just about a decade ago, *Homo Deus* (a book which was translated into 65 languages, has sold 45 million copies and was cited in over 5,000 academic publications) was introduced with the bold assertion that:

> War is obsolete. You are more likely to commit suicide than be killed in conflict. Famine is disappearing. You are at more risk of obesity than starvation. Death is just a technical problem. Equality is out—but immortality is in. (Harari, 2015)

Gaza, just under 80 km away from Prof. Yuval Noah Harari's university, has amply demonstrated that even though equality is definitely out, mortality is still very much in. It has also proven that no medical, biotech, or cyborg technology can save humanity from the unimaginable cruelty some people can exhibit towards those who cannot defend themselves. Moreover, the end of history, which Francis Fukuyama (Fukuyama, 1992) insisted was ushered by the end of the Cold War, the dissolution of the last of the 'Evil Empires' (Reagan, 1983), the Soviet Union, and the rise of western liberal democracy, did not come with the turn of the twentieth century. If anything, it is now further away than it ever was.

NATO (the North Atlantic Treaty Organization) membership has expanded from 12 to 32 countries. Six of them, namely Norway, Finland, Estonia, Latvia, Lithuania, and Poland, have a combined border of 2,557 kms with Russia thus forming what seems to be the 21st Century Iron Curtain separating 'the west' from the 30 'emerging economies' and the 'Global South' (including one NATO member, Türkiye) which were represented in the 2024 Kazan meeting of the BRICS (acronym for the founding states, namely Brazil, Russia, India, China, and South Africa) (Jutten, 2024). Türkiye's rhetoric, as well as its policies in the Balkans and its military interventions in Syria, Iraq, and Libya, are constantly challenging the Treaty of Lausanne and all the other treaties which dismantled the Ottoman Empire. Its political leaders boast that they will do what Omar, Saladin, and Sultan Selim did before them to build the Empire: after Damascus they will conquer Jerusalem (Newsroom, 2024a). Are we at the point Umberto Eco described as a history of backwards steps; "a history tired by the leaps of the past two millennia, spinning around itself, returning to the convenient greatness of Tradition" (Eco, 2006, p. 15)?

Any observant citizen of the EU, the USA, or Australia has already experienced what political scientists and philosophers, including Francis Fukuyama himself, have been stressing for at least a decade now: the western world is fast sinking towards extreme authoritarianism (Fukuyama, 2021). In the name of fighting various 'axes of evil', 'terrorists', and 'tyrants' government policies are leading us to widespread and systematic erosion of both substantive and typical democracy. This means that not only the efficiency of democratic processes and the quality of their outcomes—such as delivering justice, guaranteeing equity, equality, and fundamental rights, and making a positive difference in people's lives—is diminishing, but also that the number of countries that have democratic processes and follow them correctly is decreasing (Fukuyama, 2014). Free flow of information, ideas, people, products, and capital is just another fantasy that reality has shattered as politically driven obstacles are raised right, left, and centre (Fig. 1.2).

Fig. 1.2 Politically driven obstacles are raised right, left, and centre. Computer artwork by Anna K. Zarkada

1.1.1 Key Management Challenges Posed by Geopolitical Shifts

World War III and even a nuclear holocaust (Mecklin, 2024; Nichols, 2022) seem to be real possibilities (Stone & Pamuk, 2024) even to people who are as conservative as bankers (Blake, 2024). There are, of course, sectors, businesses, and individuals that will see their profits rising beyond belief as global conflict levels increase. For most managers, however, the level of uncertainty these imminent threats introduce to their business planning is fast becoming unmanageable. Supply chains and market access are disrupted, the cost of resources, including capital and energy, is fluctuating, and employee morale and consumer confidence are affected by the gloom and doom atmosphere. Thus, business outcomes are becoming increasingly unpredictable under the rising political and economic uncertainty.

1.1.2 Restrictions on the Access to Resources, Markets, and Market Intelligence

Crucial to how consumers', employees', and managers' attitudes and behaviours are influenced by the current PESTLE conditions prevalent in liberal economies is the shattering of the pervasive illusions of pluralism, freedom, and democracy which used to be the driving forces of twentieth-century western societies (Bock-Côté, 2022; Chomsky, 1995), the foundations of capitalism (Friedman, 1982), and the fuels of entrepreneurship (Farè et al., 2023). The curtailing of personal and civil liberties in the name of security and safety is increasingly becoming evident to even the most complacent citizens and the most absorbed in profit-making managers. Neither safety nor security are delivered, however, but the loss of freedom has wide and deep consequences at the personal, societal, and economic levels.

EU citizens, for example, have no access to Russian media because telecommunications service providers, at the behest of the president of the European Commission Ursula von der Leyen, simply block websites, apps and satellite signals (Hsu & Myers, 2024). Like Jorge of Burgos, the fourteenth-century Benedictine abbey's blind librarian who went to murderous, and ultimately self-distracting, extremes to suppress Aristotle's second book of *Poetics* on comedy because he believed that laughter outweighs the fear of God and thus undermines church authority, the EU blocks access to content that could potentially contradict its narrative. Unlike the peasants in Umberto Eco's *The Name of the Rose* (Eco, 2014[1983]), however, the modern populace are keenly aware of the censorship.

Further to the frustration people feel over the curtailing of their right to information and the disrespect shown to them by a state apparatus which considers them incapable of distinguishing between facts and propaganda, such interventions pose practical problems as well. Individuals cannot communicate with friends and relatives over blocked social media applications, businesses have their access to market and the flow of market intelligence restricted, and scholars and artists have lost access to a vast body of scientific literature and archival material. Moreover, citizens' moral evaluations of such actions negatively impact their levels of trust in the media they are allowed access to, as they know they only report one side of the news, and, what is worse, they lose confidence in once believed to be democratic institutions. This can be seen in the fact that participation in elections is falling to dangerously low levels. In the 2024 European Elections, for example, half of the voters shunned the process and in one-third of the countries participation was way lower than 40% (European Parliament, 2024).

In Greece, for example, only 41% of the people voted in the EU parliamentary elections. Given the distribution of the countries EU parliament seats, it is obvious that no more than 20% of the population actively agree with the government's stance in the Ukraine–Russia conflict. Its wide-ranging ramifications, however, hit the whole population in multiple ways. With cheap Russian natural gas left out of the country's energy mix, prices are skyrocketing, manufacturers face increased production costs, and all businesses have higher operational costs; thus, inflation rises and households become colder in the winter, hotter in the summer, and overall poorer. Food prices are also rising at rates higher than that of inflation as stock feed and fertilisers from Russia, Ukraine, and Belarus have been substituted by much more expensive imports from America. Finally, large numbers of businesses involved in trade or cross-ownership with Russia have been directly affected by the sanctions, as have shipping companies, exporters of fresh fruit and vegetables, and whole sectors like real estate and tourism. It is indicative that the once-lucrative Russian tourist market is practically lost. The numbers of Greece's highest per capita spenders dwindled from a pre-war over 600,000—800,0000 visitors to less than 15,000 in 2024 (International Newsroom, 2022; Newsroom, 2024b; Wright, 2024).

Access to resource and buyer markets is important to businesses but also critical to management is access to timely, sufficient, and reliable intelligence. 'News' like these are hardly a solid basis for decision-making.: "Russian soldiers killed: 1,351 to 43,000" (Matthews et al., 2022) or

> Bill Gates trialling Bovaer [an additive to reduce bovine enteric methane emissions], in ... milk. Bovaer is a cover for a milk-based mRNA delivery system. They plan to get every single unvaccinated person! Bo etymology means dwelling or dwelling place and vaer means man. If the holy spirit dwells within man, then attacking Gods dwelling place by attacking man (and changing his DNA) means we're no longer made in Gods image. (Raval, 2024)

Yet, the former makes it difficult for businesses with interests in Russia to estimate the time it will take for the war to end and the latter incited a social media storm and widespread consumer boycotts in the UK (Raval, 2024). Even if we accept that truth is the inescapable casualty of war and business, we expect it to find a haven in universities. But that is no longer the case either. In order to "protect students", despite the fact that they are adults, British Universities exclude from their courses the works of authors such as William Shakespeare, Geoffrey Chaucer, Jane Austen, Charlotte Morde, Charles Dickens, Agatha Christie, and August Strindberg because they contain references to the slave trade or suicide (Telegraph Reporters, 2022). YouTube™ took down a video of a roundtable discussion with scientists on managing the pandemic to prevent the propagation of 'misinformation', a move that has been described as another blatant example of big tech conglomerates trying to silence those who disagree with their corporate agenda (Siemaszko, 2021). At the same time, businesses are hard hit by repeated disinformation storms, often created by AI and mostly spreading like wildfire on social media (Raval, 2024).

Erosion of the Prerequisites of Free Enterprise: Civil Liberties, Rule of Law, and Freedom of Information

However, the above are not isolated extreme examples but rather the norm. In 2020, Freedom House recorded the lowest number of democracies with only about one in three countries assessed being a democratic one. The decline in political rights, civil liberties, and the rule of law worldwide has been steady; 2021 was the 16th consecutive year, since 1973, in which the overall democracy score fell in the 195 countries and 15 territories with a non-universally recognised state entity according to systematic measurements by independent experts (Csaky, 2022). The data these assessments are based on include issues that directly affect businesses and management practice such as the functioning of government, corruption, independence of the media, trade union rights, the rule of law, individual and property rights, and freedom of expression.

Equally grim is the picture painted by the quality of the intelligence business decisions are based on. In 2021 almost half of the countries experienced a deterioration in internet freedom (Csaky, 2022). In 2024, the world-wide-web was described as "a censored, distorted, and unreliable information space" with overall internet freedom declining for the 14th consecutive year. In more than one out of three of the 72 countries assessed, which account for 87% of the world's internet users, protection for human rights online diminished. Of the estimated 5 billion internet users, 79% live in countries where people were arrested, and even imprisoned for over 10 years, and 67% live in countries where people were physically attacked and killed for the opinions they expressed online. Finally, in more than half the countries, there have been incidents of internet and mobile networks being disconnected by the authorities for political reasons (Funk et al., 2024). It is obvious that these conditions directly threaten not only entrepreneurship in general, but also business activities, such as cause-related marketing and business communications, and even the personal safety and freedom of managers themselves and their team members.

Information integrity is also threatened by the rapidly increasing sophistication, accessibility, and affordability of generative AI which no law or regulation seems to be able to control (Funk et al., 2024; Raval, 2024). Such technologies can be used for the generation of misinformation and disinformation which can be relatively easily and cost-effectively propagated on social media to manipulate public opinion on people and brands, to influence attitudes towards products and ideas, to damage corporate reputations, and even to support lawfare which exhausts competitors' resources. The inability of governments to safeguard information flows, coupled with the documented (Funk et al., 2024) trend of big tech divesting investments in trust and safety systems, is creating a perfect storm of technological and political challenges.

Global Political Risk and the Challenges Posed by Shifting Legal Frameworks

A key characteristic of a VUCA environment is increased political risk (Zarkada, 2024), "the likelihood that political forces will cause drastic changes … [which] can affect the ownership of assets through expropriation or confiscation; … the transfer of funds … and … business operations" (Zou, 2015, p. 245). Following the politically chaotic 1970s, when the concept of political risk was first systematically studied, a key point of concurrence between experts in this, largely fragmented field (Zou, 2015), is that for societies to exploit their economic development potential it is crucial to strike the correct balance on two continua: (a) protecting domestic businesses while being open to international capital, talent, and technology transfer, and (b) guaranteeing stability without sacrificing flexibility (Akhter & Lusch, 1987). At the moment, neither one of the continua is even close to an equilibrium. Indeed, what we see is that, as the VUCA conditions settle into a permacrisis, political risk becomes hard to estimate and, subsequently, almost impossible to manage.

Globalisation, and the political tensions surrounding it, at the governmental, societal and business level have been sharply increasing political risk over the past 20 years. The process of globalisation is manifested through *inter alia* changing ownership of institutions, like banks, but also of the high streets of even small cities, the de-industrialisation of Europe, brain drain or being able to dance to K-pop and watch Bollywood movies in the other side of the world of where they were produced. The reactions to it are also evident through massive and often violent protests, the rise of separatist movements, voting for populists and nationalists in many countries, and intensified influence of consumer ethnocentrism on buying behaviour (Niros & Niros, 2024) in most countries, especially lower-income ones with high ethnic heterogeneity (Balabanis & Siamagka, 2022). What we are left with is just a long list of questions with hardly any answers in sight.

Will globalisation progress, thus further crushing local businesses and subjugating smaller economies, or will its course be reversed by sanctions against former major trading partners? At the time of writing there is a threat of protectionist measures against friends and foes alike (Palmer & Gijs, 2024). Will they be imposed, when and how? And how will the affected parties retaliate? Will there be an all-out tariff war or will the return of protectionism be sporadic and episodic? Will existing political and trade blocs clash with each other or will the biggest tensions rise within them? Whichever way it goes, political risk will continue to increase, and it will present management with unprecedented challenges. We here argue that a new approach to these challenges is urgently needed as you cannot tackle new problems with the old tools.

1.1.3 Key Management Challenges Posed by a Renegotiation of Understandings on a Global Scale

From the above discussion, it is evident that the fundamental understandings which inform long-term but also day-to-day management decisions and affect all aspects of business activity are being renegotiated on a global scale. These understandings affect managerial decisions which range from macro-scale ones, like the calculation of political risk and psychic distance (Nizielska, 2013; Stottinger & Schlegelmilch, 1998; Zarkada-Fraser & Fraser, 2002) which guide foreign trade and investment or market development decisions, to micro-environmental scale decisions such as attracting and developing talent and estimating subcultural influences on consumer preferences in the increasingly multi-lingual and multi-cultural domestic environments of previously homogeneous national spaces.

Some of the most critical currently ongoing redefinitions, if not complete renegotiations, of fundamental understandings are those related to the following:

- **Us *vs*them and the 'friend'/'foe' distinction**. We are in the midst of monumental changes in what we have considered to be known and stable in the post World War II order.
  - Yet another wave of violent redrawing of state borders in Europe and the Middle East, but also, potentially, in the Taiwan Strait.
  - A renewed contestation of the power status of states and state groupings or supranational organisations. The categorising of some countries as regional, global, or superpowers is a hot issue in many regions around the world.

Their relationships are points of friction. The same applies to numerous other countries' being, becoming or behaving as a vilayet, a protectorate, a satellite, a client, or a vassal state in relation to a larger or stronger neighbouring country, or country grouping, forcefully contesting their sovereignty and vital space.
  - The blurring and shifting of boundaries of the spheres of influence which were set by a series of treaties and agreements (e.g. Yalta, the Korean Armistice, Minsk, Abraham, etc.).
  - The redefinition of national/ethnic identity schemata and the rearrangement of the relative importance of their components (such as language, spatial origin, bloodlines, traditions, religious practice and affiliation, or shared historical references).
- **Trade blocs *vs* blocks**. Management decision-making needs to take into account ambiguities regarding the stability of institutional frameworks and the strength of the protection supranational and international organisations afford the citizens of their member states and the member states against each other and in their conflicts with non-members. For example:
  - What are the practical consequences of the ruling of the International Court of Justice (ICJ) that the Israeli occupation of Palestinian territories is against international law (Berg, 2024)? Is it legal/moral to do business in the occupied territories? Should we/Can we/Is it safe to do business with the MNCs who continue to operate in the West Bank (Brown & Tatham, 2024), and are contracts enforceable?
  - How much freedom do the four freedoms of the European single market (namely, free movement of goods, capital, labour, and the freedom to establish and provide services) allow individuals and businesses? Should the veto be abolished and how should EU countries who resist the decisions taken by the European Commission be treated?
  - How are Intellectual Property Rights (IPR) and personal data protected in different markets/regions and how easy is it to enforce GDPR where it exists?
- **Developed/developing, and First/Second/Third World,** which ties in with issues such as:
  - Stereotyping of markets, institutions, organisations or people on the basis of perceptions of the level of economic and social development of a region or country and its influence on business decisions and buyer behaviour (Zarkada-Fraser, 2001).
  - Where people can/are allowed to move/seek asylum/live and under which terms.
  - Who controls migration flows and how.
  - What are the institutional, social and emotional effects of nativism, i.e. the ideology, policy, or attitude of favouring or protecting the interests of native-born or established inhabitants against those of immigrants.

- The content and enforceability of **social contracts**, even ancient ones, such as marriage and family, and the factors determining the roles of the participants in them, such as sex and gender, race, or social class.
- The role of **institutions** and the distribution of the responsibility for their operating costs (or in the case of financial institutions, the cost of their losses and even failure), access to their services (including the pervasive problem of what constitutes essential public services and who controls public goods), and share of their profits.
- The elephant in the room: **the role of the state** and all levels of government, including the application of the principle of subsidiarity, that is the locus of decision-making being at the lowest possible hierarchical level and as closely as possible to the stakeholders most affected by the decision.

How a state defines itself and its authority (for example, if it has a constitution and if all institutions uphold it), and how well the state performs its role as a framework and overseer for all economic activity, is another hotly contested topic in numerous countries, and so we are approaching it as a separate issue in the following section.

1.2 The Nexus of the Political, Legal, and Economic Environments: Failed States

The PESTLE framework helps us organise intelligence and data and guides our thought processes. However, it masks the fact that our perceptions of polity, economy, and society are much more than just interconnected. We understand a country or market as a uniform system and not as an agglomeration of characteristics along the PESTLE dimensions. We perceive each country as a uniform framework of our individual and collective experiences, our institutionalised exchanged relationships, and our business operations. So, our attitudes and resulting decision-making are derivatives of a global evaluation.

We propose that the international literature on failed states (Acemoglu & Robinson, 2012; Chomsky, 2007; Fukuyama, 2016; Nay, 2013) can be applied to better understand the Political and Economic environments which also determine, and are delineated by, the Legal one. It goes without saying that the totally failed state—something like George Orwell's *Animal Farm* (Orwell, 2021[1944]) (Fig. 1.3)—is as much a theoretical construct as the perfectly functioning one. State function or failure is not a single score, however; it is a constellation of positions on numerous intersecting continua on which all politically defined economic spaces slide as a response to the governance decisions, actions, and reactions of their citizens. These can be summarised in the following dimensions:

- Ability and willingness of the state to
  - protect its citizens from violence or disaster,
  - provide citizens with essential and well-functioning public services,

  - maintain the exclusive exercise of power within its territory,
  - control its geographical territory and safeguard the property rights delineated by its boundaries.
- Institutions which are substantive, truly democratic (that is they do not suffer from a "democratic deficit" (Chomsky, 2007)), and inclusive rather than extractive (i.e. focused on equitable distribution of social dividends of economic activity rather than designed and operated in a way which concentrates wealth in the hands of the few at the expense of the many (Acemoglu & Robinson, 2012).
- Degree of political rot, that is, political power which has been captured by well-organised interest groups who use the system to serve their own interests at the expense of the wider public interest (Fukuyama, 2016).
- Accountability and effectiveness of the governance system and governing institutions or people.
- Active compliance with domestic and international law at all levels of polity and governance.
- Rule of law and the degree to which the elites or some organised interest groups are above or subject to the law.
- Depth and spread of corruption and criminality.
- Net quantitative and qualitative gains or losses from population movements, as in brain-drain, brain-gain, skilled and unskilled labour inflows, and refugee and immigrant flows.
- Strength, vitality, and stability of the economy.

Fig. 1.3 "The creatures outside looked from pig to man, and from man to pig, and from pig to man again; but already it was impossible to say which was which". The closing sentence of the book Animal Farm by George Orwell (Orwell, 2021[1944]). Sketch by Anna K. Zarkada

The above list highlights, yet again, that the economy and politics go hand in hand, bound by law and custom. A few aspects of the current economic environment, however, merit special attention in the context of future management and the challenges it needs to address.

1.3 Characteristics of the New Economy

In the idealistic fantasy of the peaceful end of history of the early 1990s, the spread of neoliberalism with its deregulation and privatisation, alongside rampant globalisation, were seen as the harbingers of a paradise in the shape of free market capitalism and liberal democracy which would bring unprecedented levels of material prosperity, both in industrialised countries and in poor Third World countries (Fukuyama, 1992). Aided by the technology of the 2010s, and the imminent rule of data and algorithms, they were also thought to completely eliminate hunger (Harari, 2015). That's how it looked then, at least to the optimist neoliberals. Socialists would argue that the fantasy collapsed into a rerun of the nightmare of the early twentieth-century proletariat (Fig. 1.4). What each one of us sees is a matter of perspective. Fantasy and nightmare, however, are but subjective interpretations of hard facts and numbers. And in the following sections we will discuss just that, facts and numbers.

1.3.1 Clear Signs that There is Something Seriously Wrong with Our Economic System

In 2022, in the USA, the richest country in the world, almost 40,000,000 people, i.e. more than one in 10 citizens lived below the poverty line (Oxfam, 2022). In 2023, for young adults without a high school diploma this percentage rose to over 25% and for the families of the unemployed it reached 30% (Fay, 2023). Those employed fare better, but not much. Faced with the financial uncertainty and job insecurity prevalent in the hypercompetitive labour markets, they are reported to be "turning on one another", "turning to self-destructive behaviors and beliefs", and "descent[ing] into cynicism and spectatorship"; in a nutshell, "lashing out, turning inward, tuning out, or giving up" (Putnam & Garrett, 2020, p. 8).

At about the same time, in one of the poorest parts of the world, half the Yemeni children were at risk from famine (United Nations Office for the Coordination of Humanitarian Affairs, 2021) while in the State of Palestine 45% of the population were under imminent threat of starvation and almost all of them faced food insecurity in just the first two months of the Israeli blockade and military offensive (Zagout, 2023). This, of course, is nothing new. It is history sadly repeating itself until we eventually learn the lesson so as to move on. For the time being, we are stuck.

The extreme neoliberalism that spread unchecked around the world in the 1980s and 1990s (Fukuyama, 2021) brought the world into a swamp of greed consisting

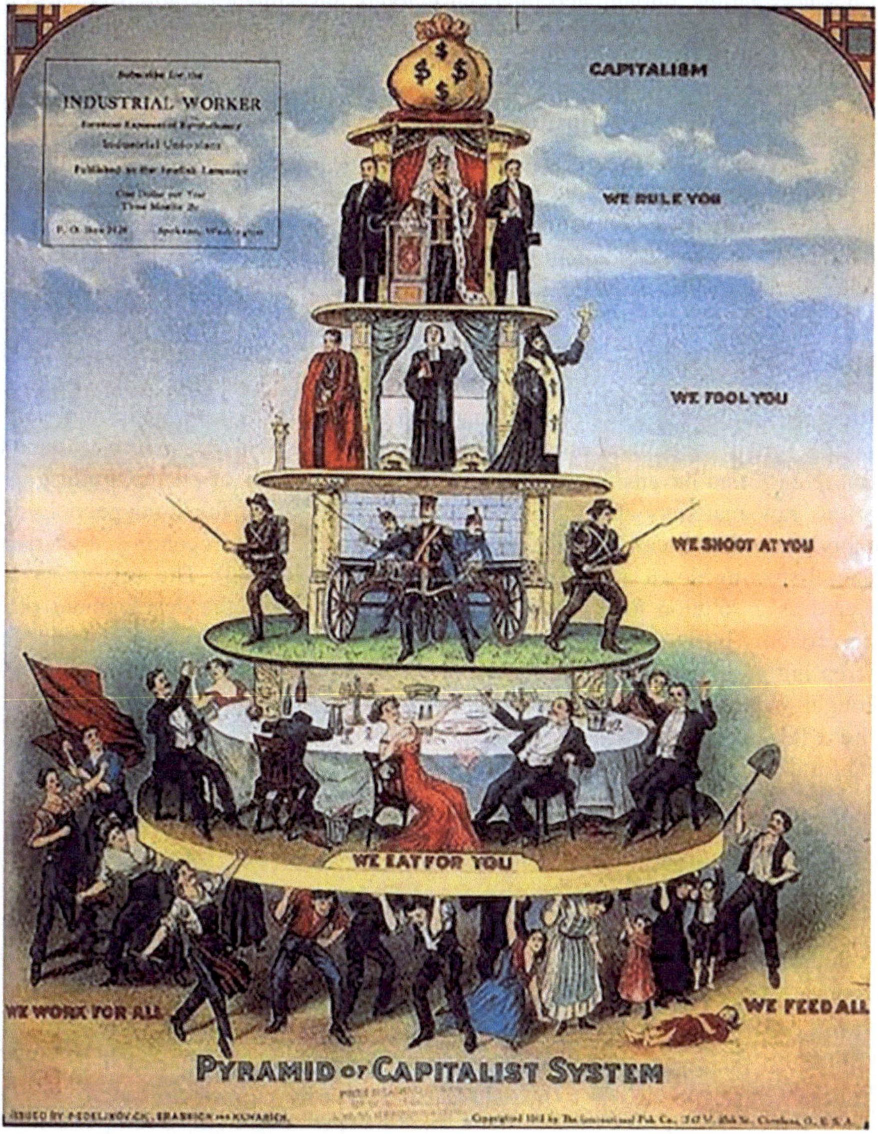

Fig. 1.4 A 1911 Industrial Worker (WWI newspaper) publication advocating industrial unionism and presenting a critique of capitalism. Available from Wikimedia Commons (https://en.wikipedia.org/wiki/File:Pyramid_of_Capitalist_System.jpg#) under a public domain licence

of business cartels, unregulated financial markets, investment pyramids collapsing one after the other, and opaque commercial practices. These practices were underpinned by an idiosyncratic capitalism where all profits of business groups went to shareholders and all losses to taxpayers. This culminated in the 2007 subprime mortgage crisis, the 2008 stock market crash, and the 2009 Eurozone sovereign debt crisis. Entrepreneurship suffered first from the erosion of free competition and then under fiscal stability programmes. Moreover, labour relations collapsed, and, even before the COVID-19 pandemic, workers were seeing their wages drop and their rights eroded (Panigyrakis & Zarkada, 2014b).

Stagflation, which seemed like a strong possibility just after the pandemic, became the frightening reality of 2024 and way beyond. It was induced by the measures taken worldwide to deal with the pandemic and intensified by the sanctions imposed on Russia as punishment for the invasion of Ukraine. The latter ended up hitting the European economy much harder than that of Russia, which found other markets for its sanctioned products, new payment methods for its cross-country transactions, and havens for its capital. Anyway, the prices of energy, grains, and various raw materials and industrial products affected by the fluid geopolitical reality are at historically high levels, which drive the world economy towards a general slowing down at fast rates (The World Bank, 2022). The only proposal offered as a solution to the economic downturn so far is the development of a war economy in Europe (Popescu, 2024). This, however, is unlikely to benefit the increasing number of people around the world who, since the winter of 2022, are brought face to face with the risk of hunger or malnutrition and poverty; people who a few short months before had an economically secure, and most of them, comfortable living (Thukral & Christina, 2022).

1.3.2 Economics 101 Fails the Permacrisis Test, Repeatedly, When It Comes to the Real Economy

Even those who can still afford life's necessities and small luxuries struggle to find products to buy. Basic products and essential raw materials which were in short supply worldwide in June 2022 (Hernandez & Wagh, 2022) are, at the time of writing, in December 2024 (Unglesbee, 2024), still scarce, overpriced, and even subject to gouging. The list is long and growing every day with no indication that the situation will improve soon. Indeed, experts expect that the shortages will affect countries all over the world for years, perhaps even more than a decade, and that the list is more likely to become longer and to expand to include items that will threaten sustenance, basic comforts, and security on a massive scale (Chambers & Packham, 2024; Dempsey, 2024; Hernandez & Wagh, 2022; Unglesbee, 2024). Some of the most indicative examples and the chain of events that caused these markets to fail are the following:

- "War and weather have put the broader food system on alert, where it may remain for the foreseeable future" as agricultural commodities are at the top of the list of severe supply chain risks (Unglesbee, 2024). The situation affects millions of businesses and billions of people worldwide. Some examples of severe shortages and resulting price hikes are:

- Cocoa, sugar, tomatoes, and olive oil because of extreme weather in West Africa, Southeast Asia, India, and Southern Europe. Prices have risen to the point where both the food processing industry and consumer groups pressure governments for direct interventions, albeit with no results.
- Grain as millions of acres of grain-producing land in Ukraine have been abandoned, 2024 yield was the third worst in over 40 years (Young, 2024), and out of season rainfall made it harder for farmers to harvest their wheat in Northwest Europe and Scotland.
- Foods containing hot peppers as a seasoning or preservative due to the prolonged drought in Mexico and California.
- Potatoes, due to floods in the producing countries. Fast food chains, such as McDonald's™, suspended the sale of large portions to avert the catastrophe of fries' complete withdrawal from their menus and whole communities have seen a significant rise in their cost of living as potatoes are the basis of their diet.
- Cream cheese due to a cyberattack on the largest producer, industry-wide labour shortages, and disruptions in the supply chain.
- Infant milk powder, the availability of which has fallen to less than half of the pre-COVID-19 levels due to the disruption of the supply chain, makes it difficult to procure raw materials and distribute the products to the points of sale.

- Timber. From April 2020 to April 2021 its price went up by 375% due to increased demand, as many people decided to renovate their homes and renew their furniture during the COVID-19 pandemic lockdowns, and reduced supply, due to port and trade route closures. Although in 2023 prices returned to late 2020 prices, the trend is, again, upward.
- Toilet paper, tissues, and paper towels. It is not only the timber shortages that reduce supply. The pandemic panic buying was repeated in 2024 because of port strikes.
- Personal hygiene products for women and disposable diapers as both cotton and plastic are scarce. Their price is on a continuous upward trend and is also dragging along the price of substitute products.
- Garage doors and other electronic and electrical systems used in home construction and repair due to shortages in circuit boards and skilled craftsmen. Together with the shortage of timber and delays in the delivery of plumbing and electrical materials, they cause an almost global delay in the completion of new housing (Badger, 2022). Add to the mixture inflation in the European countries and the USA, and you get increased house prices and shortages in the housing market. The expansion of the short-term rental of urban dwellings as vacation homes also limits supply and pushes rents up. This burdens family budgets and creates major unrest in major tourist destinations like Barcelona.

- Microchips because of raw materials supply chain disruptions. This creates shortages and dramatic price increases of PCs, mobile devices, navigation systems—anything that is connected to the internet or has any form of automation such as heated car seats and remote controls.
- Battery minerals. Demand for lithium tripled between 2017 and 2022 (Unglesbee, 2024) and is expected to quadruple again by 2030 (Statista, 2024). Supply, however, is forecasted to cover about half of the demand. The situation is similar for the rest of the minerals like cobalt and nickel. Further to the soaring prices of raw materials, finished products, and the products the batteries are used in, it seems that the greening of transportation, with the use of electric cars and busses, and the substitution of other fossil fuel-based energy sources might subsequently slow down significantly.
- Motorcycles and cars. The markets for new vehicles are disrupted by circuit board and microchip shortages, which increase the half-finished product stockpiles, trade route disruptions, and inflation. Furthermore, disruptions in the replacement parts markets make resale prices higher than those of new vehicles, thus further hitting the industry with insecurity, volatility, unpredictability, and, ultimately, bottom-line squeeze.
- Bicycles as they were purchased in large numbers during the COVID-19 pandemic lockdown, and there is still a serious instability in the supply and production lines of bicycle parts in China.
- Energy, including nuclear power as uranium production will be slashed in 2025 due to resource delivery delays and shortages (Dempsey, 2024). What is expected is local perfect storms (Chambers & Packham, 2024) and an overall VUCA global energy market (International Energy Agency, 2024). And this was before Israel launched its attack on the Iranian energy infrastructure.

The list of shortages could, unfortunately, go on and on. The grim snapshot of the real economy it shows is not likely to improve within the usual business planning time-horizon. Shortages, which create further shortages, and price hikes in numerous other industries and countries are not the only problems, however. There are also surpluses which have similarly catastrophic effects. The German car manufacturer, VW™, for example, needs to limit production by 700,000 cars in the next five years (Amann, 2014) as the estimate, in September 2024, was that demand could return to 2016 levels as late as 2033 (Business Monitor International (BMI), 2024). The data on which the forecast was drawn had not factored in Donald Trump's return to the White House. The German automotive industry, however, is not an isolated casualty of the permacrisis.

What really matters for most people and businesses, however, is the real economy, i.e. the total flow of value in society which is realised through the total production and purchase of real products (tangible goods and services excluding the global financial elements of the economy). Bitcoin might be up by 135% year-on-year, but it is the efficient market function of the real economy which determines quality of life for people and creates the basis of success for businesses.

It is, in a nutshell, the mechanism which gives us our 'daily bread', some creature comforts, and the means to pay for them. The market mechanism functions within a framework of freedom, choice, and the information needed for rational decision-making regarding resource allocation. When the framework collapses, we see shortages, surpluses, and price volatility; we see the clear signs of market failure.

Market failure, despite its being "as old as economics itself" (Marciano & Medema, 2015), is hard to define, explain and correct. However, its consequences are acutely and widely felt. What everybody experiences is the market's inability to serve its purpose, that is, to ensure the flow of value, either completely or "in an economically optimal manner" (Marciano & Medema, 2015). Moreover, prices do not reflect supply and demand and, so, the imbalance results in under- or over-production; both shortages and surpluses. What is not evident to everybody are the obstructions to efficient transactions (Dean & McMullen, 2002). These obstructions can be random and uncontrollable events, like the closure of major trade routes because of war, natural disasters, and other factors which impede the smooth running of supply chains like the ones we discussed above. They can also be caused by excessive concentrations of market power, such as monopolies or cartels which corrupt the price/supply/demand equilibrium. Usually, they are the outcome of numerous factors at play. One factor, which is always part of the market failure problem, however, is imperfect information. In essence, the obstructed flow of information causes a pervasive imbalance between marginal social costs of production and marginal social benefits and so, a few market actors benefit immensely at the expense of others and the majority (Dean & McMullen, 2002; Jackson & Jabbie, 2019; Marciano & Medema, 2015).

1.4 The Social Environment: Managing the New Organisation

Since the time of Henri Fayol (Fayol, 1917), managers are trained to use data and market intelligence for Planning, Organizing, Commanding, Coordinating and Controlling (POCCC) resources to produce the desired outcomes. Commanding and controlling might have been softened and expanded to leading and evaluating, functions might have been added, removed and questioned, but the crux of the matter remains: management is a structured process which takes place within a functioning market. When the invisible hand of the market (Smith, 1761) is taking a nap in some deep pocket, and the government refrains from waking it up, lest it is seen as intervening in the cherished 'free market' (with the exception of the money markets, which governments scramble to save from themselves at their first sneeze), management is at a loss as its theoretical tools fail to predict outcomes and thus to guide action.

Managers are not trained to take the high-risk, Squid Game-like, bets that the permacrisis world and multiple market failures throw at them. They are trained to view administration of both private and public organisations as a rational, systematic, intelligence-based decision-making process of which intuition, creativity, imagination, and risk-taking are parts, not substitutes, of planning and budgeting. The shortages we discussed above, coupled with inventory pileups (Unglesbee, 2024) caused by geopolitical instability, trade route disruption, fast diminishing consumer spending, and the redirection of government spending towards a war economy (Popescu, 2024), create a lethal cocktail which threatens the bottom line of many firms and even whole industries. The pressure on managers to find ways around the obstacles, to save the bottom line, and to keep their jobs is mounting.

The most common approach, cost cutting to preserve profitability, is not working and redesigning the whole supply and value chain with resilience in mind, which is strongly advocated by industry experts (Unglesbee, 2024), is both costly and hard to implement, and almost impossible for most firms. Coming back to the VW™ example, the negotiations between the representatives of the key stakeholders lasted for months. The last round "continued deep into the night" in "a dated no-frills business hotel", "with negotiators only taking short breaks to sleep and fuel up on coffee, curried sausage and fruit" (Amann, 2014). The "70 h of gruelling negotiations" over five days resulted in what the union leaders "hailed … as a 'Christmas miracle'": the workforce will be reduced by one quarter by 2030. Still, this is "likely not enough and over a too longer time frame to address the current stagnation" (Amann, 2014).

Even if the layoffs deliver the desired outcome for the company, the fact remains that 35,000 people will be out of a job in the very short time frame of five years, albeit "in a socially responsible manner" (Amann, 2014). Humane slaughtering does not make cows any less dead, however. It is obvious that VW™ is not an isolated example. Indeed, Forbes painfully summarises the 2025 US job market in three sentences: "(a) the unemployment rate will rise; (b) job creation will hit a sustained slump; (c) crackdown on immigrants will not create jobs and will put upward pressure on inflation" (Amdur, 2024). The forecast for the EU is similar. There is a lethal cocktail of (i) labour market failure (widespread shortages in many industries, from construction to healthcare and from ICT to agriculture, and an unmanageable surplus of highly, and expensively, educated people in others like print media, conventional manufacturing, energy, retail, and administrative support), (ii) low employee satisfaction because of systematically "declining working conditions and wages", and (iii) high emigration propensity (Gauret, 2024) and brain drain from the south to the north of Europe and from Europe to the Global South. While there is clearly a pervasive "global war for talent" (Ekkehard & Feist, 2024), there is conflicting evidence regarding the impact of immigration on labour market imbalances (Handel, 2024). The result is a huge political controversy and widespread social tensions regarding immigration and sovereign nations' response to it.

Social tensions are also caused by the increasing numbers of both the educated youth who enter the precariat of the gig economy (often glorified as digital nomadism, entrepreneurialism, and flexibility (Mancinelli, 2020)) and of those of their less privileged peers who become dispensable cogs of the platform business model. These social tensions are brought into the workplace as the managers of the twenty-first century must deal with multiple forms of employment and an increasingly multi-cultural and multigenerational workforce with diverse motivations, needs, and aspirations. Social tensions also influence buyer behaviour in unprecedented ways and directly challenge both B2B and B2C/G strategies and practices.

The generational dimension is of particular pertinence to management and marketing practice. Yet, managers are not adequately trained to deal with the fact that this is the first time in history when six generations are actively living, working, playing, and shopping together. The currently coexisting—and vying for resources, power, attention, and social services and protection—generational cohorts are the following (Beresford Research, 2023; Dimock, 2019; Jiang, 2018):

1. The affluent, extraordinarily large in numbers and surprisingly active compared with their predecessors, retirees who were born before or during World War II, referred to as the Builders or Traditionalists or Silent Generation.

---

2. The Baby Boomers (born immediately after the war and until the mid-1960s) and
3. Gen X (which ends around 1980)

   The digital immigrants who hold the key leadership positions in industry and politics.

---

4. The Millennials or Gen Y (were born in the digital world but before the end of the twentieth century) and
5. Gen Z or iGen (born in the first two decades of the twenty-first century)

   The digital natives who constitute the majority of the global workforce and the largest consumer market.

---

6. Generation Alpha, the children of the Millennials, who, although they do not have an income yet, as they were born after 2010, spend much more than children of any other generation on fictional characters, merchandising, and branded content and exert an outstanding influence over the consumption decisions of the households they live in compared to any other generation of children before them.

Even though globalisation seems to have brought about high levels of homogeneity of tastes and fashion, especially in the industrialised world, there is significant intergenerational heterogeneity in the use of mass media and information and entertainment consumption patterns. Moreover, despite the significant similarities in the education managers receive across countries and regions around the world, motivation, aspirations, and work-life balance expectations vary between generations (Beauchamp & Barnes, 2015; Kotler et al., 2021; Lee et al., 2021; Mullen, 2020; Nartea et al., 2019; Zarkada, 2021). These differences are intensified by the huge digital divide splitting the generations and by the extreme polarisation of wealth and poverty, with today's youth being the first generations who are systematically worse off than their parents and grandparents.

1.4.1 Social Tensions Challenge Management Orthodoxy

Over the past century or so, people in business organisations resort to management theory for guidance in the conduct of their organisational lives and business practices. It seems, however, that the permacrisis is rendering significant parts of what we think we know obsolete. For example, the benefits of offshoring and outsourcing are seriously challenged by trade route closures and will be further put to the test by the return of tariffs, increased protectionism through non-tariff barriers, and "the growing influence of consumer ethnocentric tendencies" (Baber et al., 2024). Similarly, branding seems rather pointless in the face of product shortages and stagflationary pressures throughout Europe which cause a continuous shrinking of consumer disposable income. In this permacrisis-dominated context the effect of branding on consumer behaviour is hard to assess at any given point in time and unpredictable in the face of shrinking incomes and rampant nationalism.

Theories of environmental factors analysis[1] and most of twentieth century thought regarding the sources of competitive and comparative advantage (i.e. the ability to outperform the competition and managing to leverage lower costs) are also in need of serious re-examination when resource and product markets are failing or, at best, showing serious imbalances of unpredictable valance. Even twenty-first-century approaches to strategy formulation are hard to implement when there are too many unknowns and change is unstructured and often violent. A case in point is the Blue Ocean Strategy (Kim & Mauborgne, 2014), the planned shift from traditional competition-focused strategies (the red ocean of cut-throat competition for a larger share of a steady or even shrinking market) to the blue oceans of uncontested market space which the firm creates to render competition irrelevant. In times of affluence and relative stability companies have both the resources and the space and time to opt for value innovation. They can focus on their customer perceptions of value and use product and service innovation to maximise their value offering. Value innovation, however, "occurs only when companies align innovation with utility, price, and cost positions" (Kim & Mauborgne, 2014). What is the cost position of a manufacturer in an energy-intensive industry when Israel is bombing Iran energy facilities and Iran is considering a move to close the straits of Hormuz (Kimball, 2025)? How can any firm align utility and price when oil prices do not just rise but they jump from day to day and they have

[1] We will discuss the most commonly used ones such as the PEST-type analyses, and Porter's diamond of national advantage and the 5 forces model for the competitive environment in [Chap. 3 Value-Values].

been doing so for over three years now (Kimball, 2025)? For as long as the permacrisis remains the largest competitor for any firm in any industry our existing market analysis and strategy planning tools will remain irrelevant.

It is not, however, just a matter of abstract entities, such as firms, facing a turbulent and almost hostile geopolitical and economic environment. Firms are nothing but groups of people and people watch the news with dread. Across industries and countries, executives and front-line employees, public sector managers and small business owners alike, are faced with job insecurity, the collapse of already shrunk social protection systems, and a rising cost of living as well as the overhanging threats to their personal safety and that of their families. The EU has issued guidelines for creating and maintaining a 72-h survival kit comprising *inter alia* water, energy bars, matches, radio, cash (after years of the Union pushing hard for a cashless economy), and official documents in a water-resistant folder. How safe can people feel while preparing their "population self-sufficiency" (European Commission, 2025) backpack? It follows that households are restructuring their budgets to hedge for the uncertainties and revert to hiding cash in the freezer or under the mattress, but certainly away from the banking system and consumption or private investment. Consequently, businesses are hard hit by the shrinking of disposable incomes and the credit crunch and resort to cost cutting and layoffs, thus causing even more shrinking of their share of a thinning wallet. Indeed, industry analysts forecast a landslide of scaling down of operations, a reverse of the globalisation process, and increased localisation and reshoring (Unglesbee, 2024). So, managers find themselves and their firms caught in a vicious cycle.

Issues that were faced by MNCs are now spread throughout the economy. Large corporations had the means to support theory development or at least purchase expert advice to address the challenges of a temporarily VUCA environment, but SMEs never did. They copied as much as they could, improvised, and plodded on as best as the environment afforded them. The more stable the conditions, the better they managed their resources and the closer they got to achieve their objectives. Nowadays, however, developments are too fast even for large corporations to be able to follow, let alone lead or determine. Despite its being strongly criticised as dehumanising workplace relationships, and even societies in their entirety, McDonaldisation (Ritzer, 2013), for example, did address some of the challenges of a sporadically VUCA environment with considerable success. Systems based on calculability and predictability which were designed to deliver high efficiency, standardisation, and total control served the mass, diverse, and relatively stable markets of the late twentieth and early twenty-first-century quite well. In a permacrisis, however, such systems are unlikely to deliver the expected results as calculability and predictability are hard to achieve and the factors to control are too complex.

Our theories need to be re-examined because they are premised on the fundamental assumption that markets are functioning efficiently within a free enterprise system. They presuppose that:

a. Consumers are sovereign and free to choose. This is a clearly unreasonable expectation when there are large-scale shortages but also when "giant technology companies.. have used their roots in the internet to influence the way we shop, consume information … search online" (McCabe, 2024) and vote (Houssiau et al., 2023), or when providers of essential services "use their power to squash would-be rivals before they're even out of the gate" (Fitzpatrick, 2014).
b. Private property is protected. This is hardly the case when households and businesses alike are debt-ridden and people are losing their jobs, thus putting their mortgaged homes at risk of repossession.
c. Profit is an incentive available to all with no prejudicial discrimination, which is impossible when markets are dominated by cartels and monopolists (Fitzpatrick, 2014; McCabe, 2024).

In failing markets, the key assumption of microeconomic efficiency, that allocation of resources maximises the total societal welfare (Jackson & Jabbie, 2019), simply does not hold. So, we need new tools to understand and navigate the new economy, and managers need to be retrained to cope with the new pressures. Coping with the permacrisis, however, is not a matter of enduring 'gruelling negotiations', sleep deprivation, and being underfed and overcaffeinated, as suggested by the VW™ negotiations imagery (Amann, 2014), which, incidentally but not coincidentally, we think, resurrects the successful perception management recipe of the "*[t]he 17-h thriller at the Summit*" (Newsroom, 2015) of the 2015 Eurogroup negotiations of the Greek bail-out (ABC News, 2015; Newsroom, 2017). It follows that, in both cases bad news had to be delivered, and in both cases market failures were at the root of the problem.

Households and businesses suffer from market failure in a way similar to that of countries. Many advanced economies are currently standing on the precipice of a sequel of 2008. Even the most illustrious neoliberal experiment, the Eurozone, is at peril as both Germany and France enter 2025 without a budget, in a stagflation, and with shaky governments (Elliott, 2024). Japan's budget also seems to be heading for a gridlock amidst political turmoil; a situation that also describes the other side of the Sea of Japan (or East Sea, if you look at it from the Korean shore). China's government is as stable as it could be but its "collapsing housing market and rising deflationary pressure" leaves the international business community holding its breath over not only the size and structure of the impending economic stimulus package, which will affect capital movement globally, but also over the form and force of the retaliations to the USA tariffs threat, which is expected to be anything but symmetrical (Kajiwara & Piccoli, 2024).

When multiple resource and product markets in multiple countries, in various stages of economic development all over the world, not only fail at once, but also remain incapable of reaching an equilibrium over an extended period, then it is neither Adam Smith, nor COVID-19, Vladimir Putin, Iran, and Hamas who are to blame. Market failure is the result of a systemic fault in the way the individuals comprising the regulatory institutions perceive their roles and responsibilities. It is also a result of a systemic deficiency in the mindset that guides the major market players in identifying and prioritising their stakeholders' rights. Finally, the lack of resistance to injustice and the excessive tolerance of the widening gap between the 'have' and 'have-nots' exhibited by the minor market players who cannot individually influence large-scale events (e.g. small firms, apathetic voters or abstainers, and amoralistic or irrational consumers) act as enablers of market failure, and ultimately, total system failure.

The New Economy is not sustainable, and neither is the current way of doing business and managing stakeholder relationships. Both are the products of a specific mindset: the current thinking on economic growth, competitiveness, and wealth creation and distribution. We discuss their shortcomings in the following section and in the next chapter we introduce sustainable prosperity for all. We analyse its foundations and dimensions and propose that it should be the overarching aim of all human activity.

1.5 The Current Approach to Development Is Not Sustainable

In a society where income is perfectly equally distributed among all its members, the Gini coefficient is zero. If only one individual possesses all the income the society produces, the Gini coefficient is 100. Both are theoretical situations. The reality of the turn of the twentieth and the beginning of the twenty-first century is shown in Fig. 1.5: we live in a world of gross inequality. There are high inequalities between countries, the inequalities within countries are increasing, and overall inequality remains at unsustainably high levels (Christensen & Lelourec, 2023). According to the latest World Inequality Report (Chancel et al., 2022), wealth is even more unequally distributed than income: The poorest 50% of the world's population own just 2% of total net wealth, while the richest 10% of people own 76% of the wealth. What is even worse, is that for every basket of basic goods a poor person can buy (Purchase Parity Power (PPP) of $4,100 per adult of the poorest 50% of the world's population in 2021), a rich one can buy 18,800 such baskets (PPP of $771,300 of the 10%). Such are the macro-level results of the neoclassical economic model of development: dismal.

It is no accident that modern economics is known since the mid-nineteenth century as the Dismal Science (Carlyle, 1849). Economies and societies which are organised on the belief that resources are scarce inevitably produce struggles over who controls supply and how to regulate demand. Struggles, in turn, inevitably produce inequalities, and inequalities end up in more struggles. Capitalism promised to "do away with anything that … could be called poverty" (Schumpeter, 2013[1942], p. 66) and end all struggles through perpetual growth. This, as Schumpeter predicted in the summer of 1935, has clearly not materialised despite the total dominance—at least in what we call 'the West'—of neoclassical economics.

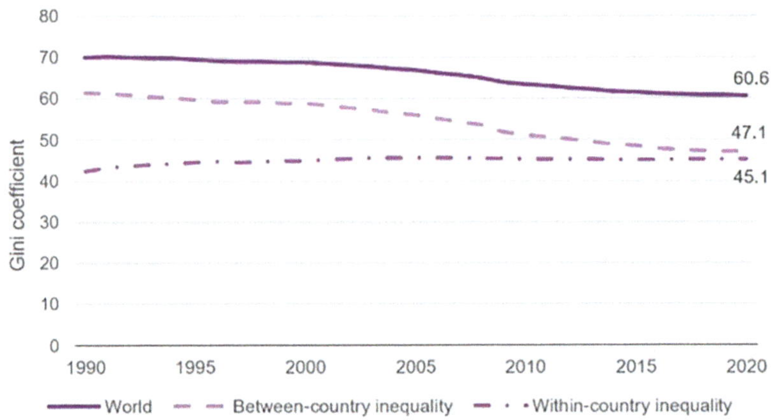

Fig. 1.5 Global inequality 1990–2020 measured by the Gini coefficient. *Data Source* Development Initiatives based on UNU-WIDER, World Income Inequality Database (WIID) Companion dataset (wiidglobal). Version 30 June, 2022 (Christensen & Lelourec, 2023), Creative Commons Attribution BY-NC-ND 4.0 International license

1.5.1 The Underlying Causes of Permacrisis: Invisible Hands Distributing Pies and Other Dangerous Fallacies

"Neoclassicism resembles racism: while ever present and dominant, no one claims to be guided by it" (Arnsperger & Varoufakis, 2006). Same as with any other prejudice, to leave it behind, we need to understand it; its origins, its function and its consequences must be laid out in the open. We here propose that it is actually the zealotic attachment to the fundamental axioms of individualism (i.e. that it is individual agents' action which causes socioeconomic phenomena), instrumentalism (that is, that all actions are means for preference-satisfaction), and equilibration (in other words the expectation that individual agents' actions will eventually stabilise in an equilibrium) (Arnsperger & Varoufakis, 2006) that have guided management theory and practice into the dead end they now find themselves: the permacrisis of the twenty-first century. We first discuss the controversies and shortcomings of current views on sustainability and prosperity in order to expose the prejudices we need to leave behind if we are to progress towards sustainable prosperity for all.

Neoclassical Economic theory, one of the ideological and methodological bases of contemporary management theory and practice, is based on the scarcity principle and sees economic activity—that is all mechanisms of ownership, production, and exchange—as the result of an imbalance between limited resource availability and unlimited human needs and desires which economic activity seeks to redress. The invisible hand of the market (Smith, 1761) regulates resource allocation through prices so that supply and demand equilibrium occurs. It follows that the fundamental understanding in our current economic organisation is that economic activity is a constant battle between people, groups, and organisations who try to obtain access to as many resources as possible at the expense of all others. The more resources the agents in a market have access to, the more they engage in transactional relationships with other actors in a never-ending struggle to maximise their well-being through the maximisation of their total utility. Their utility is manifested through ownership of as many objects as they can acquire, and it is as high as the objects are considered as valuable. Essentially, prosperity is defined as maximising the total value one achieves through transactions which make the best use of the resources at one's disposal (Matthaei, 1984).

After the Great War failed to prove itself as the war to end all wars (Wells, 2016[1914]), and while the geopolitical power structures resulting from World War II are being challenged on multiple physical, economic, and ideological warfronts, one idea seems to remain largely stable: human welfare, well-being, and happiness is equated with the acquisition of material wealth (Cassiers, 2015) and its perpetual increase (Jackson, 2013). The fixation with material possessions is not only a characteristic of the elites, whose interests it finely serves. The dream of finding happiness through ownership and consumption is prevalent among many people who are most likely to lose the wealth race, and who often do lose their health, families, and even their very livelihoods, in the process of trying to get rich. The individual and collective pursuits of people, organisations, and states have been limited to increasing wealth and profits, at the individual level, and Gross Domestic Product (GDP) (Jackson, 2015; Jain & Chhabra, 2022; Stiglitz, 2020) at the collective level.

It follows that 'development' has been the global mantra behind every economic activity with the emphasis on its quantitative meaning—i.e. growth in terms of volume and scale, a synonym of magnification and swelling. This approach encompasses a disregard of the qualitative dimension of the meaning of development—that is, the evolution towards something more capable of making a positive contribution to the world, like a child developing into an adult or a plant flowering beautifully and bearing fruit. This view of development, however, overlooks a fundamental law of nature: everything has a limit and an end. Living beings are born, grow, decline, and die. Empires emerge, expand, shrink, and disintegrate. Products have their own lifecycle which ends in their being superseded. Organisations are formed, grow, and eventually reach the limit to their ability to generate value, become irrelevant, and go into dissolution. In the context of capitalism and its concerns, the finite nature of all growth manifests as the Schumpeterian "perennial gale of creative destruction" (Schumpeter, 2013[1942], p. 82).

1.5.2 The Role of Management in the Neoclassical Economic System: Leading Ourselves into the Permacrisis We Have Created

Management was incepted, developed to its maturity, and flourished in the neoclassical economics mindset and became its key instrument. The power it holds over people and their lives has reached the point of being the dominant approach to not only the way businesses but also the way governments and societies are meant to be conducted. Public expectations that a successful businessperson will make a good statesperson are widespread in contemporary liberal democracies. Moreover, methodologies of cost–benefit analysis are considered acceptable in most capitalist economies even when they are applied to the management of public goods. So, before we examine what we can do, it is imperative to take a critical look at what we are doing which does not seem to be working as we thought it would.

Because we accept that we are doomed to fight over scarce resources in pursuit of a happiness associated with success in meeting objectives and measured in terms of possession of material objects or monetary value, we think of organisations as armies in battle over resources, markets, our time and effort, our attention, and our hard-earned money. It is, then, natural that the language of management is fraught with terms borrowed from military studies. Anyway, a whole sub-discipline of it, strategic management, has grown straight out of the battlefields of history, literally. The origin of the term 'strategy' is the ancient Greek verb *στρατηγέω* which means to lead an army to battle. Strategic management and business strategy textbooks often start with overviews of famous battles, from Marathon to Waterloo to El Alamein and reference to Alexander the Great, Sun Tzu and Clausewitz (Wunder, 2023), and indeed, it is hard to find any strategic management or marketing textbook with no references to armies, wars, and battles (Dobson et al., 2009; Lynch, 2021; Sengupta, 2011).

The interface between society and organisations, the most visible of all management activities, is marketing. Its role in the current socioeconomic structure is of paramount importance. In the context of neoclassical economic thinking, marketing, either as a discipline studying markets and the exchange function, or as a practical management activity moving value through the production and distribution mechanisms, has been the visible end of Adam Smith's hand which facilitates the reallocation of resources, the shaping of perceptions of value, and even the formation of values.

During its own classical era, the purpose of marketing was to ensure that each segment of the market would buy the product they need or want, at the price they are willing to pay, after being informed in the way and through the communication networks that suit them, through the distribution networks they have access to. Since its beginnings, marketing has seen its role as a sword which cuts through the savagery of commerce and markets (van Waterschoot & Van den Bulte, 1992). Today, and despite the widespread rhetoric about Corporate Social Responsibility (CSR) and sustainability, marketing consultants continue to advocate marketing warfare as the only way to achieve an organisation's goals with graphic passages such as:

> Marketing is war, anyway you look at it. You are battling the enemy, your competition. You are battling to dominate the market. Marketing is combat, and for every combat you need a strategy. Tighten your helmet as [the author] guides you through the trenches. (Mullen, 2009)

> "As you already know, marketing warfare is a marathon. What do I mean by that? Two things. One, you're not going to win the war overnight, and when you do take over the

> industry, you're not going to want to rest on your laurels—this fight never ends. Two, the Battle of Marathon is one of the most epic battle stories of all time (yes, when you run a 'marathon' you're reenacting the most famous event from that battle), and as you know from your school years, most battles are named after their location" (Börner, 2020).

> In an especially extreme example of a shock and awe tactic, Mongol armies catapulted human heads into enemy villages, to terrify residents. … [W]e aren't suggesting you hurl severed heads at your enemies …[b]ut the shock and awe tactic is not so dissimilar from using a concentration of resources and visibility to overwhelm the market with your brand. … This intent to undermine the enemy's morale, is a tactic that leading brands use regularly. … [O]ne big push can do significant damage to not only a competing brand's visibility, but also to their ambition. … [T]he example of Genghis Khan having his soldiers carry three lit torches to make his army look more substantial than it was is comparable to the efforts of small, new brands attempting to make themselves look more established. (Dare, 2022)

Such admonitions should not be dismissed as aberrant voices, or random business consultants trying to attract the attention of potential customers by positioning themselves as opinion leaders, albeit extreme ones. Even the academic community, admittedly in less chilling tones, has expressed similar ideas in books (Duro & Sandstrom, 1987; McNeilly & McNeilly, 2012; Ries & Trout, 1986; Storlie, 2015), and journal articles (Kotler & Singh, 1981; Taylor et al., 1990). Using the metaphor of war to describe the function of marketing perpetuates a logic of economic activity as the relentless pursuit of one's profitability at the expense of another's. In some cases, such as when a price war strategy is chosen, the profitability of one, the one that can sell at prices below cost, is based on the bankruptcy of the others, the economically weaker competitors. This, in the neoclassical liberal paradigm, is considered moral and is accepted as yet another externality of the exchange function in a free market. Business warfare is thus positioned as normal, irrespective of the social costs of market dominance and business failures, and regardless of the consequences suffered by individuals in their professional, family, and personal lives.

On a macro-scale, the operation of businesses, and even whole economies such as the EU or USA, is managed as if the scarcity constraint was the single most important consideration; an ever present and unavoidable one. We know, however, and we have so for almost a century now, that this is not the case (Matthaei, 1984; Robinson, 1969; Simon, 1979). Resource scarcity, the belief that the sum of all resources is fixed and must be divided among all economic agents, like a pie a family shares for dinner, is just a fallacy aptly called 'the fixed pie fallacy' (Kemp, 2007).

The fixed pie fallacy is a fundamental one, because "[m]ost economic fallacies derive" from it, as Milton Friedman is quoted to have stated (Fullani, 2012). And it is a fallacy at all levels. An increase in the standard of living of one country does not reduce the corresponding indicators of another. One could argue that it will actually increase the overall prosperity of neighbouring countries, as there will be no uninvited migration from the poorer to the most prosperous ones. Nor is it necessary to reduce the wealth of one person to increase that of another. My friends' increased affluence is likely to result in my getting better birthday presents from them and will certainly not restrict my access to job opportunities.

The fixed pie is also a fallacy when it comes to the allocation of responsibilities. Accountabilities are not a fixed amount shared among those who participated in producing an outcome, be it good or bad (Kaiserman, 2021). For the degradation of the natural environment, for example, each individual agent's responsibility does not tend to zero simply because it is caused by the actions of billions of individual agents (Kaiserman, 2021). Likewise, the fact that, as Senator Bernie Sanders said back in 1974, "the rich are getting richer, and the poor are getting poorer" (Follett, 2015) is by no means a natural and inevitable phenomenon. It is simply evidence that the invisible hand of the market (Smith, 1761) has failed in its role as a regulator of modern markets probably because the global, immaterial, and automated financial system operates differently from the eighteenth-century Scottish banks which Adam Smith studied.

A case in point: Banks and the painful consequences of their faulty assumptions

> "The financial sector of the U.S. economy was ballooning, driving up GDP. As it turned out, many of the profits that gave that sector such heft were, in a sense, phony. Bankers' lending practices had generated a real-estate bubble that had artificially enhanced profits—and, with their pay being linked to profits, had increased their bonuses. In the ideal free-market economy, an increase in profits is supposed to reflect an increase in societal well-being, but the bankers' takings put the lie to that notion. Much of their profits resulted from making others worse off, such as when they engaged in abusive credit-card practices or manipulated LIBOR (for London InterBank Offered Rate of interest for international banks lending to one another) to enhance their earnings.
>
> [...] GDP figures took [the] inflated figures at face value, convincing policy makers that the best way to grow the economy was to remove any remaining regulations that constrained the finance sector. Long-standing prohibitions on usury—charging outrageous interest rates to take advantage of the unwary—were stripped away. In 2000 the so-called Commodity Modernization Act was passed. It was designed to ensure that derivatives (risky financial products that played a big role in bringing down the financial system just eight years later) would never be regulated. In 2005 a bankruptcy law made it more difficult for those having trouble paying their bills to discharge their debts—making it almost impossible for those with student loans to do so (Stiglitz, 2020)".

What Joseph Stiglitz, the recipient of the Nobel Memorial Prize in Economic Sciences, wrote in 2020 could have been describing not only the US but many other economies, if not the whole of the 'western' world. The fundamental assumptions of individualism, instrumentalism, and equilibration failed to hold in the face of excessive concentration of power, including through lobbying and outright political influence, in the hands of a small number of central bankers and major international and local systemic banks. Running a business, or as is the case with banks, a whole sector, under a war-like logic results in war-like consequences. The Greek debt crisis, for example, is a case in point. The country's banking system collapse cost numerous lives because of suicides and serious diseases being undertreated due to public health system underfunding. It has burdened the economy more than World War II and the Nazi occupation of the country combined, and it has degraded human rights—individual, political, economic, and social—to alarming degrees. Austerity measures have not only affected work, education, and health but they have led to a curtailment of the right to strike and to express dissent through peacefully demonstrating against the government (Crespi et al., 2014).

The gradual liberalisation of the world's banking system, with its subsequent relaxation of internal and external restrictions, has led to the introduction of private sector management and modern marketing practices in the banking and finance industry. It has also allowed the banks to focus on increasing shareholder value and returns at the expense of fulfilling their role of value providers and supporters of the smooth operation of the real economy. Purely capitalist practices replaced the traditional state-like governance structures that were prevalent in financial institutions up until the 1980s. This led to the development of the banks within a growing rentier economy, through speculating on the markets (Jacoby, 2023) and exponential private credit growth. Both approaches have been linked to banking crises, as has liberalisation (Amri et al., 2011), but the practices spread regardless.

Operating under the neoclassical economics logic, and with an instrumentation aimed at grabbing the largest possible share of the market, aggressive bank marketing exploited the large information asymmetries between them and their customers and, for decades, engaged in blatantly abusive practices such as promoting credit cards to financially vulnerable consumers and insupportable loans to poorly managed businesses. Banks have the expertise and tools to make more accurate assessments of the creditworthiness and general financial status of their potential customers than individuals and small businesses themselves, especially those with low levels of financial literacy. Nevertheless, instead of protecting them from buying higher-risk investment packages than they could afford or taking out loans they couldn't repay, they aggressively promoted products that increased the bank's profits to the detriment of their clients. Thus, individuals, households, and businesses were reduced to the so-called red or bad loans and lost their homes and livelihoods.

The problem created by banks' retail marketing practices was so great that it led many countries, such as Cyprus (Υπουργείο Ενέργειας Εμπορίου και Βιομηχανίας (Ministry of Energy Commerce and Industry), 2021), to impose restrictive measures on the advertising of financial products, even if one could argue that it was too little too late. For example, it is a legal requirement, since 2021, that advertising should safeguard that its messages do not influence consumer attitudes through vagueness, exaggeration, or omission. Moreover, it is mandatory that all advertisements include warnings, such as "If you do not keep up with your loan repayment plan, you could lose your home", in bold type, within a frame and in a font of at least the same size as the dominant font size of the advertising copy. Even in television advertising, the warning must run throughout the duration of the advertisement.

Moreover, banks' abusive stakeholder relationship management practices, harmed society either by harming millions of consumers and investors or by causing significant distortions of the real economy. An example of such practices was the huge LIBOR scandal which resulted in shocks to many economies, some of which proved to be near-fatal ones. Banks submitted false data on estimated interbank lending rates and thereby manipulated the market to increase their profits at the expense of national economies and citizens on both sides of the Atlantic (Stiglitz, 2020; Ευρωπαϊκή Επιτροπή (European Commission), 2012).

Finally, banks, supported by shadow banking operations (Jacoby, 2023), brokerage firms, and various 'experts', 'consultants', and, malicious or foolish, rumourmongers took advantage of people's sciolism and greed and manipulated them towards irrational herd behaviours. In Greece, they managed to involve over 1,500,000 households (Πολυχρονιάδης (Polychroniadis), 2019; Χαροκόπος (Harokopos), 2021), roughly half the population, in the bubble of the Athens Stock Exchange which burst in 1999 and caused the ruin of many small investors. It is estimated that over €140,000,000,000, roughly 182% of the country's GDP, changed hands and ended up in the pockets of sharks and corrupt politicians (Πολυχρονιάδης (Polychroniadis), 2019; Χαροκόπος (Harokopos)). The then Finance Minister, Yiannos Papantoniou, boasted that "we … are the guarantors of the smooth running of the markets" (Χαροκόπος (Harokopos), 2021) and his party, the socialist party (PASOK), run the 1999 European election with the main slogan "1,000,000 Greek investors, KNOW that shares have value". This slogan was the headline in sandwich posters on a lamp posts all over the country. The bottom boxes in the poster summarised the whole scam (for which ex-minister Papantoniou and his wife were imprisoned but later released on a technicality). There was a blue one, representing the conservative party (Nea Democratia) which read "NO to instability, to experiments and to adventures" and a green one representing PASOK reading "YES to stability, to security, to development".

Development was used as an excuse for defrauding the gullible many and siphoning their savings and credit worthiness into the coffers of the unscrupulous few who knew when to jump in and when to leave the sinking boat. Stability and security were masking the dream of easy riches for the masses. Gambling was presented as the preferred alternative to skills, hard work, and professional effort. So, people jumped on the bandwagon and parted with every last cent they had, and all they could borrow from eager bankers lining their own pockets with hefty commissions on loan sales, in anticipation of grabbing a piece of the + 1,000% return that was realised by many shares of companies that appeared out of the blue, only to disappear in no time (Χαροκόπος (Harokopos, 2021). Banks, the mass media, which were in deep debt to the banks and entangled with the political elites, intermediaries, and politicians, colluded to build a fantasy world of a democratic capitalist paradise which turned into a collective nightmare with no end in sight.

The Athens Stock Exchange scandal paved the way for the 2009 combined Bank and Sovereign Debt Crisis which resulted in the over 15-year-long austerity and subsequent recession and stagflation, the irreversible impoverishment of the majority of the population, and the country being the first ever developed country to be demoted to the category of emerging economy (Peacock, 2013). It could be argued that here is a textbook case of a disorderly shakeout: it was sudden and chaotic, it was also extremely painful for far too many people as literally millions of the weaker participants in multiple markets were not only forced out of the financial markets but also of their homes and the country as, having lost everything, hundreds of thousands sought employment in other countries. In theory, the shakeout would have resulted in market corrections but all it did was to cause a violent restructuring of many sectors and a massive change of ownership of factors of production from the many to the very few. Unlike normal shakeouts, it was not intense competition leading to consolidation but corruption and a perverse form of crony capitalism that staged the shakeout.

The excessive profits of the very few were realised at the expense of the livelihood of the many. The repercussions were widely felt and affected seemingly irrelevant aspects of quality of life such as the air quality of urban centres and the rural and forest ecosystems. Impoverished city-dwellers could not afford to run their central heating systems, so they secretly cut down trees in parks to use as firewood or burnt cardboard and old furniture in makeshift fireplaces which caused a lethal fog and many poisoned families. Government services reduction also meant limited funding for forest fire prevention and management, so large ecosystems were first destroyed by uncontrollably raging bush fires all over the country and then finished off by floods caused by the lack of trees to hold the rain off arable land and rural habitats. Thus, people and the environment were cruelly sacrificed on the altar of profit.

1.6 The Current Approach to Sustainability Is Not Sustainable

> The triple bottom line is a business concept that states firms should commit to measuring their social and environmental impact—in addition to their financial performance—rather than solely focusing on generating profit, or the standard "bottom line. (Miller, 2023[2020])

It is no more than common sense that you cannot separate economic activity from its consequences on society and the natural environment. And no sane person would cherish the thought of enjoying one's wealth on a barren planet strewn with dead bodies. Yet, even in the digital hallowed halls of Harvard University, the world's leading "billionaire making machine" (Desjardins, 2017), the need to convince the business community to measure the social and environmental impacts of profit-making is acutely felt; so much so that a 2020 blog post (Miller, 2023[2020]) on why the 3BL is important was updated and reposted on 2023. The seeking of the triple bottom line (3BL—Fig. 1.6) means considering the effects of business decisions on People, Planet, and Profits (3Ps). The argument is that it makes perfect business sense to take the People (all stakeholders as opposed to the traditional focus on just shareholders) and the Planet (to the rapid demise of which "businesses have historically been the greatest contributors" (Miller, 2023[2020])) into account in the pursuit of Profits (the uncontested *raison d'être* of business).

All that is required of managers is just to "commit to measuring their social and environmental impact". Yet, it takes some effort to convince an apparently

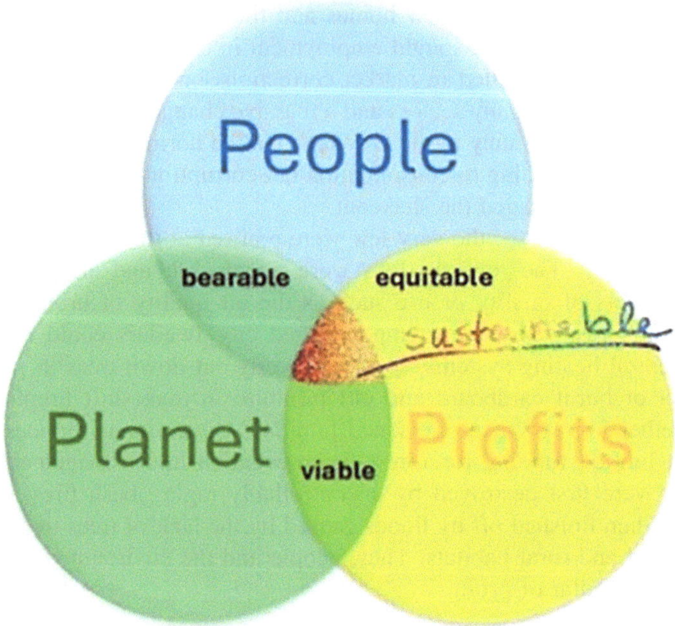

Fig. 1.6 The Triple Bottom Line (3BL). Computer artwork by Anna K. Zarkada

reluctant audience to make even such a small deviation from the relentless pursuit of profit at all costs to everybody and everything. The post (Miller, 2023[2020]) presents purpose-driven leadership as "fashionable" (Gavin, 2019), and promises that the adoption of sustainability will prove to be a driver of business success. It is not just pundits, however, that were too slow to realise the harsh realities of minding your own business and too timid to remind business leaders that they are not the sole owners of the planet. Global leaders were also way too slow to accept responsibility for the consequences of their policies on ecosystems and communities and too circumspect in suggesting some action to reverse the damage.

As late as 2016, the United Nations adopted an agenda for achieving an equilibrium of the three Ps (Miller, 2023[2020]) (people, planet, and profits) in the form of 17 Sustainable Development Goals (SDGs) to be met by 2030. The progress assessment of 2024 showed that out of a total of 169 targets, only 17% are on track, progress on 18% has stalled, and on another 17% the world has regressed below the 2015 base level (United Nations Inter-Agency and Expert Group on SDG Indicators (IAEG-SDGs), 2024). Overall, we are "severely off track" and sustainability is "[a] promise in peril" (United Nations Inter-Agency and Expert Group on SDG Indicators (IAEG-SDGs), 2024). In the words of António Guterres, the Secretary-General of the United Nations: "[t]he *Sustainable Development Goals Report 2024* makes for sobering reading" (United Nations Inter-Agency and Expert Group on SDG Indicators (IAEG-SDGs), 2024) as it makes it abundantly clear that the resolutions, pacts, signatures, and collective (in?)action of the main global governance body are consistently failing to deliver the 3BL and find the common ground among:

- The need for business enterprises to generate profits. This is positioned as the *sine qua non* for innovation and the continued supply of products and services necessary to sustain human life and to deliver prosperity. This, however, is only true in the context of capitalism, and only because the tenets of dominant variants of the neoclassical and (neo)liberal economic theory preclude state provision of even the most essential public goods. When the COVID-19 pandemic hit, however, it was the research funded by states in public universities that provided the science behind the vaccines (Cross et al., 2021) which made a proper windfall of profits for big pharma.
- The inherent and inalienable, indivisible and interdependent, universal, human rights—"from the most fundamental—the right to life—to those that make life worth living, such as the rights to food, education, work, health, and liberty" (The Office of the High Commissioner for Human Rights, 2024).
- The protection of the natural environment, climate, and life on land and below water, and their preservation for future generations.

1.6.1 The Unsustainability of the 3BL Approach to Sustainability

The common area on the 3BL Venn diagram (Fig. 1.6) is designated as 'sustainable'. According to the Oxford English Dictionary, sustainable means "able to continue over a period of time", that is, not to become extinct any time soon. The word does not imply any improvement of the quality of life of people or any reversing of the damages to the natural environment. Yet, for human existence on earth to become sustainable—which, it is not at the moment—and for the planet to become sustainable—that is, not to be destroyed—all exchange relationships need to take into account the sustainability of all three parts, all three of the 3Ps.

It is not enough to try and keep the planet and its ecosystems viable, that is to prevent it from dying an untimely death or to just keep it functioning, albeit barely. And it would make little sense to make people even poorer in the process, especially if they could not enjoy the benefits of a really cleaner and safer world. The European Green Deal seems to be doing just this at the time of writing. EU citizens bear the high cost of reducing emissions of greenhouse gases while increasing the profits of a handful of energy moguls with no hope in sight of getting cleaner air to breathe. And what good would a more equitable distribution of profits do if the environmental externalities were too high for the environment to remain viable? It would result in a richer populace with no access to clean air, water, and nature, and no safe, natural, food to enjoy. Finally, not taking into account profits goes against the very grain of the current economic system and would require a total mindset regeneration not only of the masses but also of the ruling elites. And this is not likely to happen soon enough. Nevertheless, it is important to examine the potential alternatives to businesses just minding their own business.

1.6.2 The Unsustainable Consequences of Pursuing Dual Bottom Lines, or Trying to Be Seen to Be Doing It

A curtailing of pursuing profit as the single goal by considering the environmental consequences of economic activity could make the planet viable but it would leave people behind so, it would make little sense for humans, with the exception of shareholders who live far away from their production facilities and do not use their products. Yet not even viability is really pursued. Instead, what we see is extensive and widespread greenwashing, i.e. the creation of a false perception of a product as environmentally friendly when it is not really ecological. One of the most egregious examples of such a practice was the more than 11,000,000 Volkswagen™ cars (many Porsche™, Audi™ and VW™ models) which had built-in illegal hardware and software which gave false emissions readings during inspections. So, while the Unique Sales Proposition (USP) of these cars was the low emissions which made them, supposedly, environmentally friendly, their engines emitted up to 40 times the permitted limit for nitrogen oxide pollutants (Ewing, 2017; Hotten, 2015). Well-meaning consumers paid a price premium in the belief that they were helping make the planet, and their cities, more habitable but, what they really ended up doing was to increase illegal profits for industry.

Balancing profits with human rights could make wealth distribution equitable, but this is also more of an exercise in paying lip service than an actual consideration for most business. Instead, excessive blue washing is observed in all aspects of economic activity. Blue washing takes its name from the colour of the UN flag and describes the concealment of the negative social effects of the activities of businesses which, although they have signed the United Nations Global Compact, have not really incorporated the principles of the 17 UN Sustainable Development Goals (UNSDGs) into their practices. One such example is Nestlé™ which, despite it's proudly demonstrating the logo of being a signatory to the pact, has been accused of using cocoa harvested by unpaid children in Africa (Balch, 2021). Eco-conscious European and American consumers feel at peace with themselves for participating in recycling programmes, yet it is the cottage industry of recycling of batteries that is causing decreased intelligence, aggressiveness, and a host of other developmental and health problems in millions of children in the DRC (Murray, 2021). E-waste (Marshall, 2021) and fast fashion discards (May, 2019) tell similar stories of environmental racism towards many millions of people in various African countries.

Racism, however, comes in many other forms and shapes and its flames are fanned by seemingly innocent business practices which are so widespread and enmeshed in consumer markets that they simply go unnoticed. Yet, practices like whitewashing, which are very common in the advertising and entertainment industries, undermine human rights in the most insidious way. The term whitewashing refers to the practice of making the skin colour of non-Caucasian models, artists, or celebrity endorsers look whiter than it really is with the use of photographic filters, cosmetics, or digital manipulation of photographs. This way, the celebrities, and subsequently the products they are associated with, are supposed to become more attractive to audiences drawn to the European standard of beauty. A classic whitewashing example is the L'Oréal™ print advertisement in which the famous singer Beyoncé looked almost white when using a hair colour that "shimmers from the inside out" (Amri et al., 2011; Sweney, 2008). Whitewashing, further to its being insulting to the racial identity of billions of people, also causes low self-esteem in young people all over the world and feeds the multi-billion-dollar skin-whitening and cosmetic surgery industry.

Sacrificing profits to ensure the welfare of people and the protection of the natural environment, which could make life on earth bearable, is presented as an impossibility in the capitalist system. An outstanding range of abusive business practices such as the ones described above results in unsustainably large imbalances between people, planet, and profits as, instead of offering the consumer a total utility equal to the utility of the seller without burdening the environment, they increase the profits of the seller disproportionately to the value offered to the other parties in the exchange relationship while leaving the negative externalities (such as pollution, or competitors' failure, for example) to be borne by the planet and society. This imbalance is caused by a religious-like devotion, if not obsession, with profit maximisation and economic growth.

1.6.3 The Painful Consequences of Putting Profits and Growth Over People and the Planet

The single-minded pursuit of profitability and economic growth is such a skewed approach to prosperity that even New Keynesian economists such as Joseph Stiglitz consider the GDP to be an imperfect and ineffective measure of well-being (Stiglitz, 2020). Even as a metric, the GDP does not reflect a country's progress, not even the size of its economy, as it measures only the volume of money exchanged for goods and services with no consideration for the value of unpaid domestic work and the monetary equivalent of caring for children, the sick, or elderly family members, subsistence agriculture, and the voluntary sector. Moreover, it does not factor in the costs of environmental degradation or social inequality (Moore et al., 2021) and it ignores the quality of social relationships, financial and personal security, health and longevity (Jain & Chhabra, 2022).

As it turns out, the burgeoning GDP and hyper-profits of giant business conglomerates have neither increased the life satisfaction of the people of developed countries, nor defeated global poverty (Cassiers, 2015)—they have not even achieved a fairer distribution of wealth within the advanced economies (as shown in Fig. 1.5). Moreover, the negative environmental externalities are piling up. In recent years, and especially since the outbreak of the COVID-19 pandemic, more and more scientists, intellectuals, and citizens are realising that human civilisation has reached a turning point in history where the strain on the planet's natural limits seems to lead to catastrophic consequences such as irreversible damage to the natural environment and the earth's climate. Environmental and social problems, as well as successive and simultaneous nutritional, energy, economic, political, and cultural crises are interrelated phenomena related to this narrow conception of well-being (Zarkada, 2024).

Humanity's trajectory has, for about a century now, been viewed as constant growth. Economic growth was expected to translate into higher incomes for all, which were, in turn, expected to bring more and better choices, richer and longer lives, improved quality of life, and an automatically achieved prosperity equated with having more of everything. But things didn't turn out that way. At the individual level, not only does development not bring happiness to people, but as relatively recent research has shown, "economic growth and human misery go hand in hand" (Shankar et al., 2006, p. 487). In practice, as the economy expands, so do its impacts on the resources associated with it (Jackson, 2015). These impacts are already unsustainable. In the last 25 years of the twentieth century, the global economy doubled but 60% of the world's ecosystems were degraded (Jackson, 2013). The effects on people are no less severe.

> Obesity, diabetes, and poor cardiac health come not from insufficiency but from over- sufficiency. The diseases of affluence are a symptom of incoherent thinking. Even in something as basic as food, the equation of more with better just doesn't stack up. Beyond this simple example, it is not difficult to show that it isn't so much materiality itself that brings us affluence but the capabilities that we have to flourish as human beings. (Jackson, 2015, p. xiii)

Death by Lifestyle

The 'diseases of affluence' are not just a matter of individual bad luck or poor dietary choices. They are the symptoms of the painful, contagious, and socially transmitted condition caused by the endless pursuit of material accumulation known as affluenza (De Graaf et al., 2005). Affluenza (a word play of affluence and influenza) manifests as 'having a lifestyle' (Panigyrakis & Zarkada, 2014a) of flaunting conspicuous leisure and consumption (Veblen, 1998[1899]) aimed at short-term bonding with other people (Sundie et al., 2010). Even though affluenza is classless, when the rich are growing richer and the poor poorer (Follett, 2015), affluenza delivers double punches to the latter. First, it brainwashes them into desiring a life full of shiny trinkets, and then it hits them with the realisation that they are given minimal opportunity to obtain them (De Graaf et al., 2005).

It follows that affluenza is linked to debt, anxiety, and waste at the household level while at the societal level it is related to problems such as poverty, worker and human rights violations, destruction of natural habitats, and pollution. Thus, we end up living in inhumane urban environments inhabited by isolated people suffering from depression, obesity, and overwhelmingly unserviceable debt. We need to keep in mind, however, that affluenza is not a force of nature but a direct result of putting profits and growth over people and the planet. It is embedded in the ideological foundation of the current economic development model which promotes overconsumption. It also has its sponsors and apologists.

Mainstream marketing, with its superficial interpretation of the Freudian concept of desire (Schmitt et al., 2021; Solomon et al., 2006), propagates an almost religious belief in the inevitability of the endless cycle of desire → acquisition → new desire as the key—if not sole—motivator and purpose of human existence. These beliefs are expressed with hollow phrases like 'the customer is king' or 'customer-centric [organisational] culture' (Shankar et al., 2006; Woodall, 2012) and underpin ideologically driven grand theories like Service-Dominant (S-D) Logic (Vargo & Lusch, 2008, 2018), which underplay the role of the employees in value co-production, overplay the individual over the collective, and cover under the flattening term 'consumer/s' the essence of the human existence expressed, for example, by terms such as 'citizen' and 'person/s'.

The financially, psychologically, socially, and environmentally harmful promotion of overconsumption to increase profits is so deeply embedded in management literature and practice that we no longer notice it; just as fish do not notice the water they swim in. It also guides the development of theory, like consumer culture theory (Arnould & Thompson, 2005; Hungara & Nobre, 2021), which presents the disease—affluenza—as the cure for existential angst and a cornerstone of identity construction (Panigyrakis & Zarkada, 2014a). Thus, we see a glorification of overconsumption as retail therapy (Rick et al., 2014) and the overall promotion of consumption as so much more than what it really is: a solution to mundane problems such as protecting your body from the cold or getting from A to B.

It is not just the obese that die of affluenza. Social cohesion also disintegrates by the widening gap between the under- and the over-privileged. The rich do not enjoy their riches because they see others having so much more and the poor who watch *Sex and the City* or *The Real Housewives* experience their poverty as more severe than it truly is (Csikszentmihalyi, 1999; Woodall, 2012). At the end of the day, managers find themselves in a position to try to not only make people want more but also to convince them that they should want whatever they cannot have and that they must strive to acquire more so that they can have an Instagrammable lifestyle which will classify them among the accepted, modern, cosmopolitan, beautiful, desirable, etc.

A decade after the calculations of the costs of affluenza for the natural environment and human health were made (Jackson, 2013, 2015), the so-called developed world, the very small number of countries with the very large economies (International Monetary Fund, 2021), is facing not only a combined energy and food crisis while standing on the precipice of a nuclear holocaust but also its own fragility. Suddenly, the service and experience economy make little sense and what matters is water, food, and energy security. The real economy, such as agriculture and heavy industry, which was all but ignored while the world admired the rise of tech startups and glorified venture capital, futures, and derivatives is now painfully acknowledged as our only survival raft. What is real is gradually being re-understood as real and not a social construction. Understanding how we got here and what we can do now, when it is already 90" to midnight on the doomsday clock (Mecklin, 2024), is the single most pressing task we are faced with. It is a matter of survival for the human species and the prerequisite of current prosperity. It is also the basis for future substantive and equitably accessible sustainable development, as in growth, not just enlargement of numbers to a point beyond human scale.

1.7 The Fixation with 'Profit Above All' Hurts Both Businesspeople and Businesses

Businesses which are focused on growing in scope and scale, dominating ever more markets, and increasing their profits, however, are also suffering the dire consequences of their choices. Understanding business as warfare incurs not only social costs but also negative consequences to their owners, leaders, and managers. These costs are emotional, social, and organisational as, ultimately, the coveted profit bottom line suffers too.

1.7.1 The Unbearable Life of a Warrior Without Rest

The words we use colour our world, structure our most basic perceptions, and organise our thinking in a very specific way. So do the metaphors we use to convey finer dimensions of meaning. Metaphors can be full of contradictions but still create powerful connections which mean much more than a term used literally (Scruton, 2012). The use of metaphors is a fundamental mechanism of the mind which allows us to use what we know about our physical and social experience in order to understand countless other things that we do not know. Thus, metaphors ultimately guide our actions without even our realising it (Lakoff & Johnson, 1980[2008]).

If we describe business planning as strategy, draw our marketing plans as if they were battle plans (Kotler & Singh, 1981), talk about market dominance (Tellis, 2012), guerrilla marketing (Baltes & Leibing, 2008; Gkarane et al., 2019; Kotler & Singh, 1981), attack plans (Kotler & Singh, 1981), reputation wars and new-media snipers (Gaines-Ross, 2010), trade wars, marketing and advertising campaigns, capturing or losing markets and market share (Wiliński, 2017), winning or losing takeover battles, or recruiting and retrenching employees, then we construe a bleak and fiercely competitive, cut-throat world with little room for caring for people and the planet, or for nurturing mutually satisfying relationships with other human beings inside the organisation and outside it. Moreover, we invite stakeholders to enter into their relationships with organisations as if they are coming into a battle ground. We pit employees against management and customers against businesses. Finally, the polemic approach of industry and businesspeople towards society not only arms shooters of executives, but also fuels rallies of public support with protesters waving posters that read "people over profit" outside the courthouse where the shooter is tried for murder (Elassar et al., 2024).

Even if we accept that military studies do have something to teach businesses, it is debatable whether the simplistic analogies, such as customer = friend and competitor = enemy, reflect reality and help the development of theory or its practical application. Indeed, in recent years it appears that people express their politics not only with what they shop (Zarkada, 2021), but also through the brands they boycott (Edelman, 2024). Customers are turning against businesses, especially large multinationals, by exhibiting increasing cynicism, reduced loyalty and increasingly strong resistance to marketing messages (Klein, 2002, 2017; Panigyrakis & Zarkada, 2014a, 2014b; Raval, 2024; Zarkada, 2013). They do not just resist passively, however. They actively boycott businesses and use their Social Media (SM) accounts to inflict as much reputational damage as they can to businesses that do not meet their quality, service, or moral standards (Amine et al., 2005; Burns & Bruner, 2000; Carrigan & Attalla, 2001; Chatzopoulou & de Kiewiet, 2021; Klein et al., 2004; Knight et al., 2022; Schmitt et al., 2021). Marketing, being the interface between businesses, people, and society, takes the brunt of people's reaction to business practices. Socially responsible investing also directs funds away from the wrong-doers and towards those who are, or appear to be the 'good' guys, at least for as long as they remain so.

There are, of course, voices that call for a reframing of the concepts and activities of management and a replacement of the polemic mentality with one of conflict resolution, diplomacy, compromise, and the recovery of trust through placing emphasis on flexibility and the proper management of available resources (Kolar & Toporišič, 2007). Even supporters of capitalism vehemently reject the war-like metaphors as deviations from its true spirit and propose the conceptualisation of corporations as excellence-seeking communities, embedded in the societal moral fabric, and organised on the principles of integrity, honesty, fairness, and compassion (Gare & Neesham, 2022; Melé, 2012; Solomon, 1992, 1994, 2004). The refutation of the "ultimate split or antagonism between individual self-interest and the greater public good" (Solomon, 2004) as a guiding principle of business activity echoes Aristotle's virtue ethics. His student, Alexander the Great, universally recognised for his strategic prowess, was, after all, living proof of the value of focusing on building relationships of trust not only with his citizens, soldiers, and allies but, first and foremost, with the inhabitants of the territories he annexed to his empire and his former rivals (Fuller, 2004).

1.7.2 Single-Mindedly Minding Your Own Business is Bad for Business

Viewing marketing as a battle and its practitioners as ad-wielding fighters aiming to eliminate competitors and dominate the market overlooks the importance of well-trained and dedicated employees who are happy to offer solutions to life's big and small problems and derive personal value by making the everyday life of their customers better. It also overlooks customers whose loyalty, reviews, and even complaints help an organisation survive, grow, and improve. And, anyway, the history books dedicate many pages to armies which, despite having numerical and technological superiority, were defeated by people who, despite their being almost unarmed, were dedicated to their ideals. For the past three decades or so, a number of fundamentally human-centred branches of marketing, such as relationship, social, and experiential marketing, are developing rapidly alongside alternative metaphors of business as a game, or as an organism which develops and adapts to its environment, or as partnerships based on commitment and trust and developing through alliances and synergies (Rindfleisch, 1996). Even Philip Kotler has moved away from his 1980s study of marketing warfare (Kotler & Singh, 1981) to embrace Human to Human Marketing (H2H) (Kotler et al., 2020, 2023), a back to the basics approach placing human beings at the centre of marketing and branding.

Humans are, indeed, hard-wired for justice (Decety & Yoder, 2016), empathy (Decety & Yoder, 2016; Iacoboni, 2009; Stueber, 2019; Wieseke et al., 2012), and compassion (Immordino-Yang et al., 2009). Thus, viewing exchanges as zero-sum games goes against human nature and, although a prevailing one, it still is a faulty logic which will continue to degrade the environment and violate human rights if not abandoned, along with the grandiose perpetual development fantasy. Exchange relationships are the exact opposite of a zero-sum game. A transaction cannot be completed unless both parties believe that they are deriving value from it. The problem with the abusive practices often employed in business warfare is that the estimate of the value the consumer will get from the transaction is not accurate—it is overstated, or the product of outright deception. To this, eventually, people will catch on and react. It is very likely that, as global insecurity increases and purchase parity decreases on a massive scale, the number of people who will stand critically towards their trust and wealth being squandered by unscrupulous businesses will grow. It is very likely that resentment towards exploiters will continue to grow and business empires might face the fate of imperial powers of the nineteenth and early twentieth centuries (Hobsbawm, 1987). Disillusionment with institutions and their power to protect people from corporate immorality seems to also grow. It is characteristic that in 2023 less than one in four citizens of the OECD countries trusted their countries' governments, and in most EU countries their trust had shrunk from that of 2021 (Organisation for Economic Co-operation and Development (OECD), 2024). It is not at all unlikely that sooner, rather than later, people will impose an end to the 'I win—you lose' game in which they invariably stand to lose.

1.8 The Contemporary Management Mindset Causing the Permacrisis

Many academics as well as business professionals now recognise the pivotal role of people and place them at the heart of all their efforts. As humans, however, we are not only customers or employees. We are social beings who also value things which are not necessarily related to wealth such as the quality of our social relationships, our participation in society, the beauty of nature, and good health (Jain & Chhabra, 2022). COVID-19 has brutally reminded us of the limits of individualism and instrumentalism; the limits that we would have been aware of if we had paid attention to history. New diseases will always appear, and old and forgotten ones will, eventually, return. It is a fact of life. It is also a fact that global deadly pandemics are becoming increasingly severe as well as more frequent in the twentieth and twenty-first centuries (Poorolajal, 2021). As people travel and migrate in large numbers and over continents in a matter of hours, local epidemics are increasingly likely to become global pandemics in speeds that were unimaginable in the past. So, human health will increasingly rest on communication, cooperation and collective action on a large, if not truly global, scale.

Moreover, the technologies we know have in operation are shaking the foundations of equilibration and, literally, blow the fixed pie fallacy out of the water. Here are some examples:

- Fossil fuels might be finite but renewable ones are not.
- Labour, both manual and intellectual, is potentially in infinite supply too as manufacturing robots never sleep, and my AI assistant can bring up in seconds the information it would take me hours to find in a conventional library.
- Printing, in 3, 4, 5…X dimensions, makes economies of scale irrelevant, elevates product personalisation back to the level pre-industrialisation artisan-made artefacts, and reconfigures the parameters of logistics, the role of designers, and the meaning of branding.
- X-realities redefine the temporal and spatial limits of socialisation, play, and even travel and experience seeking. You can now not only ride a carriage through the cobbled streets of Joseon-era Seoul, climb the stairs of the Hermitage with Catherine the Great, and fly over the Agora of fifth-century BCE Athens but, also, interact with the mayor, civil servants, and citizens of Seoul, the Metaverse capital of the world, 24/7, 365 days a year. Thus, time and space are also fast becoming infinite if not irrelevant.
- Capital is also anything but a scarce factor of production as the world debt shows us: it was US$88,273,380,861,313 when I started writing this sentence and has already grown by US$23,546,208 and during the 40" I stayed on the world debt clock it accrued well over US$3,000,000 of interest (commodity.com, 2024). Modern capital is even more imaginary and convention-based than fiat money. It goes without saying that debt, interest, money, national and other currencies, and the all too visible cryptocurrencies, are just conventions we decide to adhere to (Davies, 2002) and, like all other conventions, we can choose to throw to the wind.

We also know, since at least 1933, that there are serious doubts about the validity of concepts such as consumer sovereignty and conceptualisations such as the supply-and-demand curve (Robinson, 1969). The list of controversies inherent in our current thinking is potentially endless but, the crux of the matter remains: the resource scarcity assumption holds little validity—if it ever had any (Kemp, 2007; Matthaei, 1984; Simon, 1979). And on these controversies rests the fundamental shortcoming of the contemporary management mindset: the perpetual war over ownership of the factors of production and the strife to increase one's consumption at the expense of that of another's is irrational and will increasingly become ever more, and then some more, irrelevant, at best, and, most likely, catastrophic.

Having explained the problems management faces in a permacrisis environment, the next chapter discusses the concept of sustainable prosperity and analyses the first pillar of what we believe is the way out of the permacrisis, that is, sustainable development with all (and not at the expense of other organisations, the environment and society). We start with a thorough discussion of the concept of sustainable prosperity which we believe should be the goal of all economic—well, really, all human—activity. We envisage sustainable prosperity as both the purpose, and the new state of being for organisations, societies, and people who have escaped the permacrisis.

The following chapters discuss how sustainable prosperity can be achieved through the implementation of Future Management, the systematic, mindful, and values- and purpose-driven employment of technology to develop novel approaches to the transformation of organisational design, business practices, corporate processes, value co-creation, and seamless phygital stakeholder relationship building and maintenance. Future Managed organisations and institutions have strategic agility built in the core of their design and management. Future Managers can escape the permacrisis trap and turn the doomsday clock back through the regeneration of their mindsets towards addressing people, planet, and their organisation as a unity to be managed delicately and respectfully in order to deliver sustainable prosperity for all.

Future Management in Practice: Lessons from the Permacrisis

This chapter's focus has been on providing a *Contextual Diagnosis.* The insights we gather from the permacrisis highlight one important lesson for leaders and executives: they no longer can assume crises can be managed or treated as temporary interruptions to a previously stable system. In the world of Future Management, uncertainty is not the exception to the rule but the rule itself. This insight demands a deeper and structural shift in how organisations are shaped, led, and assessed.

1. Rethinking the Managerial Function

Managers must evolve from a simple role of merely planning for efficiency to being architects of resilience. Adaptive systems cannot be managed by rigid and linear plans or static forecasts and require a more comprehensive approach to management. This entails the development of strategic sensing capabilities with the ability to constantly monitor political, economic, social, technological, legal, and environmental signals and decipher these into actionable strategies.

2. Decision-Making Under Uncertainty

The premise of Future Management, which no one disputes, is that decisions will be made with a lack of information. Managers should embrace iterative approaches; they should pilot, test, learn, and adjust to avoid either overconfidence or paralysis. A mindset of experimentation replaces the illusion of control.

3. Multi-Stakeholder Accountability

In a permacrisis, organisations are not isolated. Managers are expected to understand that their choices are affected by a multitude of factors and affect employees, customers, supply chains, communities, and even ecosystems. Future Management calls for a stakeholder framework that considers immediate business results and greater societal equilibrium.

4. Developing Organisational Resilience

In the context of Future Management, traditional management understanding of "bouncing back" as resilience is outdated. Resilience is understood as the capacity to withstand shocks and adapt to changes while improving. Therefore, optimising supply chains, acquiring flexible capabilities, and developing employees who embrace rather than fight change is highly strategic.

5. From Reactive to Proactive

The permacrisis does not teach us to wait for stability; stability is not coming. Future Management requires that leaders shape their surroundings: participate in policy, work with other industries, and innovate to prevent rather than manage new crises.

In Practice:
Future Management, given the circumstances of the permacrisis, means accepting volatility, designing organisations to flourish amid change, and restating the notion of success as the capability to maintain continuity and legitimacy despite persistent instability.

MANAGERS MUST ACCEPT TURBULENCE AS THE NORM, CULTIVATE STRATEGIC SENSING, AND DESIGN FOR RESILIENCE.

References

ABC News. (2015). Greece bailout: Alexis Tsipras accepts 86-billion-euro bailout from eurozone leaders after 17 hours of talks. *ABC News*. Retrieved December 31, 2024, from https://www.abc.net.au/news/2015-07-13/eurozone-leaders-say-agreement-reached-at-greece-talks/6616448?utm_campaign=abc_news_web&utm_content=link&utm_medium=content_shared&utm_source=abc_news_web

Acemoglu, D., & Robinson, J. A. (2012). *Why nations fail: The origins of power, prosperity, and poverty*. Profile Books.

Akhter, H., & Lusch, R. F. (1987). Political risk: A structural analysis. In S. T. Cavusgil (Ed.), *Advances in international marketing* (Vol. 2). JAI Press.

Amann, C. (2014). VW reaches union deal to cut 35,000 German jobs after gruelling talks. *Reuters*. https://www.reuters.com/business/autos-transportation/rapprochement-between-volkswagen-union-wage-talks-sources-say-2024-12-20/

Amdur, E. (2024). Expect these three job market reversals in 2025. *Forbes*. https://www.forbes.com/sites/eliamdur/2024/11/25/expect-these-three-job-market-reversals-in-2025/

Amine, L. S., Chao, M. C. H., & Arnold, M. J. (2005). Executive insights: Exploring the practical effects of country of origin, animosity, and price–quality issues: Two case studies of Taiwan and Acer in China. *Journal of International Marketing, 13*(2), 114–150. https://doi.org/10.1509/jimk.13.2.114.64855

Amri, P., Angkinand, A. P., & Wihlborg, C. (2011). International comparisons of bank regulation, liberalization, and banking crises. *Journal of Financial Economic Policy, 3*(4), 322–339. https://doi.org/10.1108/17576381111182909

Arnould, E. J., & Thompson, C. J. (2005). Consumer culture theory (CCT): Twenty years of research. *Journal of Consumer Research, 31*(4), 868–882.

Arnsperger, C., & Varoufakis, Y. (2006). What is neoclassical economics? The three axioms responsible for its theoretical oeuvre, practical irrelevance and thus, discursive power. *Panoeconomicus, 53*(1), 5–18.

Baber, R., Sankpal, S., Baber, P., & Gulati, C. (2024). Consumer ethnocentrism: What we learned and what we need to know?—A systematic literature review. *Cogent Business & Management, 11*(1), 2321800. https://doi.org/10.1080/23311975.2024.2321800

Badger, E. (2022, February 15). 4 bed, 3 bath, No garage door: The unlikely woes holding up home building. Supply-chain complications are giving the industry and buyers fits. *TheUpshot*. https://www.nytimes.com/2022/02/15/upshot/homes-garage-door-shortage.html

Balabanis, G., & Siamagka, N. T. (2022). A meta-analysis of consumer ethnocentrism across 57 countries. *International Journal of Research in Marketing, 39*(3), 745–763. https://doi.org/10.1016/j.ijresmar.2021.12.002

Balch, O. (2021). Mars, Nestlé and Hershey to face child slavery lawsuit in US. *The Guardian*.

Baltes, G., & Leibing, I. (2008). Guerrilla marketing for information services? *New Library World, 109*(1/2), 46–55. https://doi.org/10.1108/03074800810845994

Beauchamp, M. B., & Barnes, D. C. (2015). Delighting baby boomers and millennials: Factors that matter most. *Journal of Marketing Theory and Practice, 23*(3), 338–350. https://doi.org/10.1080/10696679.2015.1032472

Beresford Research. (2023). *Age range by generation*. https://www.beresfordresearch.com/age-range-by-generation/

Berg, R. (2024). UN top court says Israeli occupation of Palestinian territories is illegal. *BBC News*. https://www.bbc.com/news/articles/cjerjzxlpvdo

Blake, S. (2024). World war III has already begun, JP Morgan Boss Says. *Newsweek*. Retrieved November 17, 2024, from https://www.newsweek.com/world-war-three-has-already-begun-jp-morgan-ceo-says-1976854

Bock-Côté, M. (2022). *La Révolution racialiste, et autres virus idéologiques*. Place des éditeurs.

Börner, K. (2020, May 27). Marketing is warfare—Marketing is war. Part 3: Know your battlefields. *White Buffalo Creative*. https://whitebuffalocreative.com/marketing-warfare/marketing-warfare-part-3/

Brown, D., & Tatham, E. (2024). Multinationals in Europe continue to flout international law in Israel's occupied territories. *Orient XXI*. https://orientxxi.info/magazine/multinationals-in-europe-continue-to-flout-international-law-in-israel-s,7743

Brown, G., El-Erian, M., Spence, M., & Lidow, R. (2023). *Permacrisis: A plan to fix a fractured world*. Simon & Schuster UK.

Burns, J. P., & Bruner, M. S. (2000). Revisiting the theory of image restoration strategies. *Communication Quarterly, 48*(1), 27–39.

Bushby, H. (2022). Permacrisis declared Collins Dictionary word of the year. *BBC News*. Retrieved December 26, 2023, from https://www.bbc.com/news/entertainment-arts-63458467

Business Monitor International (BMI). (2024). *Germany Autos Report—Q1 2025* (Fitch Solutions Issue). https://store.fitchsolutions.com/autos/germany-autos-report

Carlyle, T. (1849). Occasional discourse on the negro question. *Fraser's Magazine for Town and Country, XL*(July to December), 670–679.

Carrigan, M., & Attalla, A. (2001). The myth of the ethical consumer—do ethics matter in purchase behaviour? *Journal of Consumer Marketing, 18*(7), 560–578. https://doi.org/10.1108/07363760110410263

Cassiers, I. (2015). *Redefining prosperity*. Routledge. https://doi.org/10.4324/9781315777955

Chambers, G., & Packham, C. (2024). 'Perfect storm' warning issued to Australia's east coast energy market. *The Weekend Australian (Online)*. Retrieved December 27, 2024, from https://www.theaustralian.com.au/nation/politics/perfect-storm-warning-issued-to-australias-east-coast-energy-market/news-story/b2afd5798fdcef12d9e41fdcb3c0582e

Chancel, L., Piketty, T., Saez, E., & Zucman, G. (2022). *World inequality report 2022*. World Inequality Lab.

Chatzopoulou, E., & de Kiewiet, A. (2021). Millennials' evaluation of corporate social responsibility: The wants and needs of the largest and most ethical generation. *Journal of Consumer Behaviour, 20*(3), 521–534.

Chomsky, N. (1995). *Necessary illusions: Thought control in democratic societies*. House of Anansi Press Incorporated.

Chomsky, N. (2007). *Failed states: The abuse of power and the assault on democracy*. Metropolitan Books.

Christensen, Z., & Lelourec, H. (2023). *Inequality: Global trends*. Retrieved May 2, 2023, from https://devinit.org/resources/inequality-global-trends/

commodity.com. (2024). *World debt clock*. Moneda Media. Retrieved August 11, 2024 from https://commodity.com/data/debt-clock/

Crespi, E., Aubry, S., Gomez, M., Thiele, B., & Sant'ana, M. (2014). *Υποβαθμίζοντας τα δικαιώματα: Το κόστος της λιτότητας στην Ελλάδα (Degrading rights: The cost of austerity on Greece)*.

Cross, S., Rho, Y., Reddy, H., Pepperrell, T., Rodgers, F., Osborne, R., Eni-Olotu, A., Banerjee, R., Wimmer, S., & Keestra, S. (2021). Who funded the research behind the Oxford–AstraZeneca COVID-19 vaccine? *BMJ Global Health, 6*(12), Article e007321. https://doi.org/10.1136/bmjgh-2021-007321

Csaky, Z. (2022). *Nations in transit 2021: The antidemocratic turn*. Freedom House.

Csikszentmihalyi, M. (1999). If we are so rich, why aren't we happy? *American Psychologist, 54*(10), 821.

Dare, M. (2022). Marketing warfare: What parallels could we possibly draw between marketing and war? *Marketing Theory*. https://murraydare.co.uk/marketing-theory/marketing-warfare

Davies, G. (2002). *A history of money: From ancient times to the present day*. University of Wales Press.

De Graaf, J., Wann, D., & Thomas, N. (2005). *Affluenza: The all-consuming epidemic*. Berrett-Koehler Publishers.

Dean, T. J., & McMullen, J. S. (2002). *Market failure and entrepreneurial opportunity academy of management proceedings*. Briarcliff Manor.

Decety, J., & Yoder, K. J. (2016). Empathy and motivation for justice: Cognitive empathy and concern, but not emotional empathy, predict sensitivity to injustice for others. *Social Neuroscience, 11*(1), 1–14. https://doi.org/10.1080/17470919.2015.1029593

Dempsey, H. (2024). World's largest uranium producer slashes production target. *FT.com*. Retrieved December 27, 2024, from https://www.ft.com/content/240af090-8684-49dc-a85e-20b535d62dda

Desjardins, J. (2017). Chart: Harvard is billionaire university. *Money*. https://www.visualcapitalist.com/chart-harvard-billionaire-university/

Dimock, M. (2019). Defining generations: Where Millennials end and Generation Z begins. *Research Topics—Generations and Age*. Retrieved September 7, 2022, from https://www.pewresearch.org/fact-tank/2019/01/17/where-millennials-end-and-generation-z-begins/

Dobson, P. W., Starkey, K., & Richards, J. (2009). *Strategic management: Issues and cases*. Wiley.

Duro, R., & Sandstrom, B. (1987). *The basic principles of marketing warfare*. Wiley.

Eco, U. (2006). *A Passo di Gambero—Guerre Calde e Populismo Mediatico*. RCS Libri Bompiani.

Eco, U. (2014[1983]). *The name of the rose*. Mariner Books.

Edelman, R. (2024). *2024 Edelman trust barometer—Special report: Brands and politics*. https://www.edelman.com/sites/g/files/aatuss191/files/2024-06/2024%20Edelman%20Trust%20Barometer%20Special%20Report%20Brands%20and%20Politics%20Final.pdf#page=1.00

Ekkehard, E., & Feist, L. (2024). Tomorrow at work: The age of shortages. *Intereconomics, 59*(3), 125–131. https://doi.org/10.2478/ie-2024-0026

Elassar, A., Tucker, E., Scannell, K., & Bailey, C. (2024). Luigi Mangione, CEO killing suspect, pleads not guilty to state terror and murder charges. *CNN US*. Retrieved January 4, 2025, from https://edition.cnn.com/2024/12/23/us/luigi-mangione-arraignment-new-york/index.html

Elliott, L. (2024). Germany and France are in crisis—Is the next global financial crash brewing? *The Guardian*. https://www.theguardian.com/commentisfree/2024/dec/19/germany-france-crisis-next-global-financial-crash

English, J. (2024, December 28). *'Not the new normal'—2024 'one of the worst years in UNICEF's history' for children in conflict* https://www.unicef.org/press-releases/not-new-normal-2024-one-worst-years-unicefs-history-children-conflict

Etzioni, A. (1988). *The moral dimension*. The Free Press.

European Commission. (2025). *EU preparedness union strategy: Getting ready to respond to crises*. Directorate-General for Communication. https://commission.europa.eu/topics/preparedness_en

European Parliament. (2024, September 6). *2024 European election results, Turnout by country (%)*. Retrieved January 4, 2025, from https://results.elections.europa.eu/en

Ewing, J. (2017). Inside VW's campaign of trickery. *New York Times*. Retrieved July 5, 2017, May 07, from https://www.nytimes.com/2017/05/06/business/inside-vws-campaign-of-trickery.html

Farè, L., Audretsch, D. B., & Dejardin, M. (2023). Does democracy foster entrepreneurship? *Small Business Economics, 61*(4), 1461–1495. https://doi.org/10.1007/s11187-023-00737-7

Fay, B. (2023, December 21). *Poverty in the United States*. Debt.org. https://www.debt.org/faqs/americans-in-debt/poverty-united-states/

Fayol, H. (1917). Administration Industrielle et Générale: Prévoyance, Organisation, Commandement, Coordination, Contrôle. In. Paris, France: Dunod et Pinat.

Fitzpatrick, A. (2014). A judge ordered Microsoft to split. Here's why it's still a single company. *Time*. Retrieved December 30, 2024, from https://time.com/3553242/microsoft-monopoly/

Follett, C. (2015). Senator sanders and the fixed pie fallacy. *Policy Commons*. https://policycommons.net/artifacts/1308143/senator-sanders-and-the-fixed-pie-fallacy/1911438/

Friedman, M. (1982). *Capitalism and freedom* (Reissue with new Preface by the Author ed.). The University of Chicago Press.

Fukuyama, F. (1992). *The end of history and the last man*. Free Press.

Fukuyama, F. (2014). *Political order and political decay: From the industrial revolution to the globalization of democracy*. Profile Books.

Fukuyama, F. (2016, December 13). *America: The failed state*. Prospect Magazine. Retrieved December 17, 2019, from https://www.prospectmagazine.co.uk/magazine/america-the-failed-state-donald-trump

Fukuyama, F. (2021). *After the end of history: Conversations with Francis Fukuyama* (M. Fasting, Ed.). Georgetown University Press.

Fullani, A. (2012). The competitiveness of the Albanian economy and a policy agenda. In *Growth in Albania and South East Europe: The Way Ahead* (pp. 57). South East European Studies at Oxford (SEESOX).

Fuller, J. F. C. (2004). *Η Ιδιοφυής Στρατηγική του Μεγάλου Αλεξάνδρου (The ingenious strategy of Alexander the Great,* K. K. Κολιόπουλος, Konstantinos, Trans.). Ποιότητα (Poiotita) Wordworth Editions.

Funk, A., Vesteinsson, K., & Baker, G. (2024). *Freedom on the Net 2024: The struggle for trust online* (Freedom House, Issue. https://freedomhouse.org/sites/default/files/2024-10/FREEDOM-ON-THE-NET-2024-DIGITAL-BOOKLET.pdf

Gaines-Ross, L. (2010). Reputation warfare. *Harvard Business Review, 88*(12), 70–76.

Gare, A., & Neesham, C. (2022). Creating an effective business ethics: Is virtue ethics up to the challenge? In C. Neesham & S. Segal (Eds.), *Handbook of philosophy of management* (pp. 793–808). Springer.

Gauret, F. (2024). Rebalancing the scales: Why is Europe's labour market in disarray? *euronews*. Retrieved December 30, 2024, from https://www.euronews.com/business/2024/02/07/balancing-the-labour-market-how-is-europe-doing

Gavin, M. (2019). HBS professor explores the impact purpose can have on your organization. *Business Insights*. https://online.hbs.edu/blog/post/how-can-purpose-impact-business-performance

Gkarane, S., Efstratios-Marinos, L., Vassiliadis, C. A., & Vassiliadis, Y. (2019). Combining traditional and digital tools in developing an international guerilla marketing strategy: The case of a SME Greek company. In A. Kavoura, E. Kefallonitis, & A. Giovanis (Eds.), *Strategic innovative marketing and tourism*. Cham.

Handel, M. J. (2024). Labor shortages: What is the problem? *Intereconomics, 59*(3), 136–142. https://doi.org/10.2478/ie-2024-0029

Harari, Y. N. (2015). *Homo Deus: A brief history of tomorrow* (e-book ed.). Vintage.

Hernandez, D., & Wagh, M. (2022). These 18 items are in short supply due to COVID-related supply chain issues. *Popular Mechanics—Culture*. https://www.popularmechanics.com/culture/g38674719/covid-shortages/

Hobsbawm, E. J. (1987). *The age of empire, 1875–1914*. Pantheon Books.

Hotten, R. (2015). Volkswagen: The scandal explained. *BBC News*. https://www.bbc.com/news/business-34324772

Houssiau, F., Sapieżyński, P., Radaelli, L., Shmueli, E., & de Montjoye, Y.-A. (2023). Detrimental network effects in privacy: A graph-theoretic model for node-based intrusions. *Patterns, 4*(1), 1–12.

Hsu, T., & Myers, S. L. (2024). Europe Banned Russia's RT network. Its content is still spreading. *The New York Times*. https://www.nytimes.com/2024/05/30/business/media/russia-rt-disinformation-europe-ban.html

Hungara, A., & Nobre, H. (2021). A consumer culture theory perspective of the marketplace: An integrative review and agenda for research. *International Journal of Consumer Studies, 45*(4), 805–823.

Iacoboni, M. (2009). Imitation, empathy, and mirror neurons. *Annual Review of Psychology, 60*(1), 653–670. https://doi.org/10.1146/annurev.psych.60.110707.163604

Immordino-Yang, M. H., McColl, A., Damasio, H., & Damasio, A. (2009). Neural correlates of admiration and compassion. *Proceedings of the National Academy of Science USA, 106*(19), 8021–8026. https://doi.org/10.1073/pnas.0810363106

International Energy Agency. (2024). *Oil 2024: Analysis and forecast to 2030*. https://share.google/NIDVXr0avA1dRea7I

International Monetary Fund. (2021). World economic outlook update. *World economic outlook reports*. https://www.imf.org/en/Publications/WEO/Issues/2021/01/26/2021-world-economic-outlook-update
International Newsroom. (2022). War in Ukraine and the "blows" on the Greek economy. *To BHMA (To Vima)/Wall Street Journal*. Retrieved December 26, 2024, from https://www.tovima.gr/2022/03/10/international/war-in-ukraine-and-the-blows-on-the-greek-economy
Jackson, E. A., & Jabbie, M. (2019). Understanding market failure in the developing country context. In W. Leal Filho, A. M. Azul, L. Brandli, P. G. Özuyar, & T. Wall (Eds.), *Decent work and economic growth* (pp. 1–10). Springer International Publishing. https://doi.org/10.1007/978-3-319-71058-7_44-1
Jackson, T. (2013). Prosperity without growth. In *Globalisation, economic transition and the environment* (pp. 105–128). Edward Elgar Publishing.
Jackson, T. (2015). Preface. In I. Cassiers (Ed.), *Redefining prosperity*. Routledge London. https://doi.org/10.4324/9781315777955
Jacoby, J. (2023, March 15). *Age of easy money*. PBS. https://www.youtube.com/watch?v=EpMLAQbSYAw
Jain, A., & Chhabra, S. (2022). Redefining Prosperity. *Stockholm+50: Unlocking a better future*. Retrieved July 24, 2022, from https://euagenda.eu/upload/publications/redefining-prosperity-stockholm50backgroundpaper.pdf
Jiang, J. (2018, May 2). *Millennials stand out for their technology use, but older generations also embrace digital life*. Pew Research Centre. Retrieved July 5, 2018, from http://www.pewresearch.org/fact-tank/2018/05/02/millennials-stand-out-for-their-technology-use-but-older-generations-also-embrace-digital-life/
Jutten, M. (2024, November 8). *Outcome of the 16th BRICS Summit in Kazan, Russia*. Think Tank European Parliament. Retrieved December 26, 2024, from https://www.europarl.europa.eu/thinktank/en/document/EPRS_ATA(2024)766243
Kaiserman, A. (2021). Responsibility and the 'Pie Fallacy.' *Philosophical Studies, 178*(11), 3597–3616. https://doi.org/10.1007/s11098-021-01616-1
Kajiwara, K., & Piccoli, W. (2024). *What to watch in 2025: Outlooks for the year ahead*. https://www.teneo.com/app/uploads/2024/12/What-to-Watch-in-2025.pdf#page=23.20
Kemp, S. (2007). Psychology and opposition to free trade. *World Trade Review, 6*(1), 25–44. https://doi.org/10.1017/S1474745606003089
Kim, W. C., & Mauborgne, R. (2014). *Blue ocean strategy* (expanded). Harvard business review Press.
Kimball, S. (2025). Oil prices jump more than 3%, adding to last week's surge, as Israel strikes Iran energy facilities. *Oil Prices and News*. https://www.cnbc.com/2025/06/15/oil-prices-jump-after-israel-strikes-energy-facilities-in-iran.html
Klein, J. G., Smith, N. C., & John, A. (2004). Why we boycott: Consumer motivations for boycott participation. *Journal of Marketing, 68*(3), 92–109.
Klein, N. (2002). *No logo* (paperback ed.). Picador.
Klein, N. (2017). *No is not enough: Defeating the new shock politics*. Penguin UK.
Knight, H., Haddoud, M. Y., & Megicks, P. (2022). Determinants of corporate sustainability message sharing on social media: A configuration approach. *Business Strategy and the Environment, 31*(2), 633–647. https://doi.org/10.1002/bse.2941
Kolar, T., & Toporišič, A. (2007). Marketing as warfare, revisited. *Marketing intelligence & planning, 25*(3), 203–216.
Kotler, P., Kartajaya, H., & Setiawan, I. (2021). *Marketing 5.0.: Technology for humanity*. Wiley.
Kotler, P., Pfoertsch, W., & Sponholz, U. (2020). *H2H marketing: The genesis of human-to-human marketing*. Springer International Publishing.
Kotler, P., Pfoertsch, W., Sponholz, U., & Haas, M. (2023). *H2H marketing: Case studies on human-to-human marketing*. Springer International Publishing.
Kotler, P., & Singh, R. (1981). Marketing warfare in the 1980s. *The Journal of Business Strategy, 1*(3), 30.

Lakoff, G., & Johnson, M. (2008). *Metaphors we live by*. University of Chicago Press. (Original work published 1980).

Lee, P., Böhm, K., & Westcott, K. (2021). Traditional TV wanes: Television is about to dip below half of all UK video viewing. *Deloitte Insights*. https://www2.deloitte.com/uk/en/insights/industry/technology/technology-media-and-telecom-predictions/2022/tv-viewership-decline.html

Lynch, R. (2021). *Strategic management*. Sage Publications.

Mancinelli, F. (2020). Digital nomads: Freedom, responsibility and the neoliberal order. *Information Technology & Tourism, 22*(3), 417–437.

Marciano, A., & Medema, S. G. (2015). Market failure in context: introduction. *History of Political Economy, 47*(suppl_1), 1–19. https://doi.org/10.1215/00182702-3130415

Marshall, M. (2021). Electronic waste in Nigeria sheds light on the ongoing problems of environmental racism. *Michigan State International Law Review*. https://www.msuilr.org/new-blog/2021/5/27/electronic-waste-in-nigeria-sheds-light-on-the-ongoing-problems-of-environmental-racism

Matthaei, J. (1984). Rethinking scarcity: Neoclassicism, neomalthusianism, and neoMarxism. *Review of Radical Political Economics, 16*(2–3), 81–94. https://doi.org/10.1177/048661348401600204

Matthews, A. L., Stiles, M., Nagorski, T., & Rood, J. (2022). The Ukraine war in data: Counting Russian casualties—in the 'fog of war'. *Grid*. Retrieved August 19, 2022, from https://www.grid.news/story/global/2022/08/18/the-ukraine-war-in-data-counting-russian-casualties-in-the-fog-of-war/

May, J. (2019). Designer tracks Europe's throwaway clothing to Salaula market in Zambia. *Twyg*. Retrieved August 14, 2024, from https://twyg.co.za/designer-tracks-europes-throwaway-clothing-to-salaula-market-in-zambia/

McCabe, D. (2024). 'Google is a monopolist,' Judge rules in landmark antitrust case. *The New York Times*. Retrieved August 21, 2024, from

McNeilly, M., & McNeilly, M. R. (2012). *Sun Tzu and the art of business: Six strategic principles for managers*. OUP USA.

Mecklin, J. (2024). A moment of historic danger: It is still 90 seconds to midnight. *Bulletin of the atomic scientists, 2024 Doomsday clock statement*. https://thebulletin.org/doomsday-clock/current-time/

Melé, D. (2012). The firm as a "community of persons": A pillar of humanistic business ethos. *Journal of Business Ethics, 106*, 89–101.

Miller, K. (2023[2020], August 16). The triple bottom line: What it is & why it's important. *Business Insights*. https://online.hbs.edu/blog/post/what-is-the-triple-bottom-line

Moore, H. L., Davies, M., Mintchev, N., & Woodcraft, S. (2021). *Prosperity in the twenty first century: Concepts, models and metrics*. Institute for Global Prosperity, UCL.

Mullen, C. (2020). *Gen Buy: Millennials projected to spend $1.4 trillion as influence grows*. The Business Journals. Retrieved May 15, 2021, from https://www.bizjournals.com/bizwomen/news/latest-news/2020/01/gen-buymillennials-projected-to-spend-1-4-trillion.html?page=all

Mullen, J. (2009). Marketing is war. *Australian Anthill Magazine* (December 2008—January 2009). Retrieved July 25, 2022, from http://anthillonline.com/marketing-is-war/#:~:text=In%20'The%20Art%20of%20War,to%20invite%20challenge%20and%20defeat.

Murray, L. (2021). Toxic charge: How batteries are poisoning Kinshasa's children. *Al Jazeera*. Retrieved August 14, 2024, from https://www.aljazeera.com/features/2021/10/29/toxic-charge-how-batteries-are-poisoning-kinshasa-children

Nartea, M., Jamice, J., & Goyenechea, A. (2019). Exploring the characteristics of Millennials in online buying behaviour. *Asia Pacific Journal of Multidisciplinary Research, 3*(4), 1–9.

Nay, O. (2013). Fragile and failed states: Critical perspectives on conceptual hybrids. *International Political Science Review, 34*(3), 326–341. https://doi.org/10.1177/0192512113480054

Newsroom. (2015). Το θρίλερ των 17 ωρών στη Σύνοδο Κορυφής (Teh 17-hour thriller at the summit. *iefimerida.gr*. Retrieved December 31, 2024, from https://www.iefimerida.gr/news/216824/thriler-ton-17-oron-sti-synodo-koryfis

Newsroom. (2017). Αποκαλύψεις Ρέντσι για την 17ωρη διαπραγμάτευση—Το «It's enough» και η εξάντληση του Τσίπρα (Renzi's Revelations of the 17-hour Negotiation—The "It's enough" and Tsipra's exhaustion. *CNN Greece.* Retrieved December 31, 2024, from https://www.cnn.gr/politiki/story/89141/apokalypseis-rentsi-gia-tin-17ori-diapragmateysi-to-it-s-enough-kai-i-exantlisi-toy-tsipra

Newsroom. (2024a). Erdogan's Ally Bahceli promises supporters Turkey will take Jerusalem. *To BHMA (To Vima) / Wall Street Journal, World.* Retrieved December 26, 2024, from https://www.tovima.com/world/bahceli-erdogans-ally-bahceli-promises-supporters-turkey-will-take-jerusalem/

Newsroom. (2024b). Greek tourism: Which nationalities spent the most in 2024. *To BHMA (To Vima) / Wall street journal, travel.* Retrieved December 26, 2024, from {Newsroom, 2024 #5770}https://www.tovima.com/travel/greek-tourism-which-nationalities-spent-the-most-in-2024/

Nichols, M. (2022). West could trigger nuclear war over Ukraine, Russia says at U.N. *Reuters—Europe.* https://www.reuters.com/world/europe/west-could-trigger-nuclear-war-over-ukraine-russia-says-un-2022-08-02/

Niros, M. I., & Niros, A. (2024). Marketing solutions to the challenges of a Volatile, Uncertain, Complex and Ambiguous (VUCA) environment: Applications in international marketing. In A. Zarkada (Ed.), *Marketing solutions to the challenges of a VUCA environment* (pp. 193–215). Springer Nature Switzerland.

Nizielska, A. (2013). The role of psychic distance in the process of companies internationalization-review of research results. *Journal of Economics and Management, 13*, 85–99.

Organisation for Economic Co-operation and Development (OECD). (2024). *Trust in government.* Organisation for Economic Co-operation and Development (OECD). Retrieved October 27, 2024, from https://www.oecd.org/en/topics/sub-issues/trust-in-government.html

Orwell, G. (2021[1944]). *Animal farm.* Diamond Books.

Oxfam. (2022). *Poverty in the USA.* Oxfam. Retrieved August 19, 2022, from https://www.oxfamamerica.org/explore/countries/united-states/poverty-in-the-us/

Palmer, D., & Gijs, C. (2024). EU warned to prepare for early Trump tariff action. *Politico.* https://www.politico.com/news/2024/11/15/european-union-trump-tariffs-00189639

Panigyrakis, G. G., & Zarkada, A. K. (2014a). New philosophical paradigms in marketing: From amoral consumerism to axiological societing. In L. Moutinho, E. Bigne, & A. K. Manrai (Eds.), *The Routledge companion on the future of marketing.* Routledge.

Panigyrakis, G. G., & Zarkada, A. K. (2014b). A philosophical investigation of the transition from Integrated Marketing Communications to metamodern meaning cocreation. *Journal of Global Scholars of Marketing Science, 24*(3), 262–278. https://doi.org/10.1080/21639159.2014.911494

Peacock, L. (2013). Greece reclassified to 'emerging market' from developed. *The Telegraph.* http://www.telegraph.co.uk/finance/economics/9904969/Greece-reclassified-to-emerging-market-from-developed.html?fb

Poorolajal, J. (2021). The global pandemics are getting more frequent and severe. *Journal of Research in Health Sciences, 21*(1), e00502–e00502. https://doi.org/10.34172/jrhs.2021.40

Popescu, N. (2024). 'Europe must build a war economy without delay'. *Le Monde.* https://www.lemonde.fr/en/opinion/article/2024/04/01/europe-must-adapt-and-build-a-war-economy-without-delay_6666993_23.html

Putnam, R. D., & Garrett, S. R. (2020). *The upswing: How America came together a century ago and how we can do it again.* Simon & Schuster.

Raval, A. (2024). The disinformation storm is now hitting companies harder. *Financial Times.* Retrieved December 26, 2024, from https://www.ft.com/content/0aa9725d-e423-4a6b-b842-866ad4541dc2

Reagan, R. (1983, March 8). "*Evil Empire*"*Speech by President Reagan—Address to the National Association of Evangelicals* [Speech]. YouTube. https://www.youtube.com/watch?v=FcSm-KAEFFA

Rick, S. I., Pereira, B., & Burson, K. A. (2014). The benefits of retail therapy: Making purchase decisions reduces residual sadness. *Journal of Consumer Psychology, 24*(3), 373–380. https://doi.org/10.1016/j.jcps.2013.12.004

Ries, A., & Trout, J. (1986). *Marketing warfare*. McGraw-Hill.

Rindfleisch, A. (1996). Marketing as warfare: Reassessing a dominant metaphor. *Business Horizons, 39*(5), 3–10.

Ritzer, G. (2013). *The McDonaldization of society*. Sage.

Robinson, J. (1969). *The economics of imperfect competition* (2nd ed.). Palgrave Macmillan. https://doi.org/10.1007/978-1-349-15320-6

Rosati, C. S. (2016). Moral motivation. In E. N. Zalta (Ed.), *The Stanford encyclopedia of philosophy* (Winter 2016 ed.).

Schmitt, B., Brakus, J. J., & Biraglia, A. (2021). Consumption ideology. *Journal of Consumer Research, 49*(1), 74–95. https://doi.org/10.1093/jcr/ucab044

Schumacher, E. F. (1973). *Small is beautiful: Economics as if people mattered.* Harper & Row.

Schumpeter, J. A. (2013[1942]). *Capitalism, socialism and democracy* (1st ed.). Routledge.

Scruton, R. (2012). *Modern philosophy: An introduction and survey*. A&C Black.

Sen, A. (1988). *On ethics and economics*. Blackwell.

Sengupta, N. K., & Chandan, J. S. (2011). *Strategic management*. Vikas Publishing House.

Shankar, A., Whittaker, J., & Fitchett, J. A. (2006). Heaven knows I'm miserable now. *Marketing Theory, 6*(4), 485–505. https://doi.org/10.1177/1470593106069933

Siemaszko, C. (2021). YouTube pulls Florida governor's video, says his panel spread Covid-19 misinformation. *NBC News*. Retrieved August 19, 2022, from https://www.nbcnews.com/news/us-news/youtube-pulls-florida-governor-s-video-says-his-panel-spread-n1263635

Simon, H. A. (1979). Rational decision making in business organizations. *The American Economic Review, 69*(4), 493–513.

Smith, A. (1761). *The theory of moral sentiments* (2nd ed.). A. Millar & A. Kincaid and J. Bell.

Solomon, M., Bamossy, G., Askegaard, S., & Hogg, M. K. (2006). *Consumer behavior* (European ed.). Pearson Education.

Solomon, R. C. (1992). *Ethics and excellence: Cooperation and integrity in business*. Oxford University Press.

Solomon, R. C. (1994). *Above the bottom line: An introduction to business ethics* (2nd ed.). Harcourt Brace College Publishers.

Solomon, R. C. (2004). Aristotle, ethics and business organizations. *Organization Studies, 25*(6), 1021–1043.

Statista. (2024). *Supply and demand of lithium worldwide in 2022 and forecast for 2030.*

Stiglitz, J. E. (2020). GDP is the wrong tool for measuring what matters. *Scientific American*. Retrieved July 25, 2022, from https://www.scientificamerican.com/article/gdp-is-the-wrong-tool-for-measuring-what-matters/

Stone, M., & Pamuk, H. (2024). Biden allows Ukraine to use US arms to strike inside Russia. *Reuters*. Retrieved November 18, 2024, from https://www.reuters.com/world/biden-lifts-ban-ukraine-using-us-arms-strike-inside-russia-2024-11-17/

Storlie, C. (2015, January 26). Military tactics that matter to marketing. *On marketing.*

Stottinger, B., & Schlegelmilch, B. B. (1998). Explaining export development through psychic distance: Enlightening or elusive? *International Marketing Review, 15*(5), 357–372.

Stueber, K. (2019). Empathy. In E. N. Zalta (Ed.), *The Stanford encyclopedia of philosophy* (Fall 2019 Edition ed.).

Sundie, J. M., Kenrick, D. T., Griskevicius, V., Tybur, J. M., Vohs, K. D., & Beal, D. J. (2010). Peacocks, porsches, and Thorstein Veblen: Conspicuous consumption as a sexual signaling system. *Journal of Personality and Social Psychology, 100*(4), 664–680. https://doi.org/10.1037/a0021669

Sweney, M. (2008). Beyoncé knowles: L'Oreal accused of 'whitening' singer in cosmetics ad. *The Guardian*. https://www.theguardian.com/media/2008/aug/08/advertising.usa

Taleb, N. N. (2010). *The Black Swan: The impact of the highly improbable* (2nd ed.). Random House Publishing Group.

Taylor, R., Coulter, R., & Coulter, M. (1990). Marketing warfare strategies: A model. *Journal of Business Strategies, 7*(2), 113–124.

Telegraph Reporters. (2022). Universities admit purge of 'challenging' books to protect students. *The Telegraph.* Retrieved August 19, 2022, from https://www.telegraph.co.uk/news/2022/08/09/universities-admit-purge-challenging-books-protect-students/

Tellis, G. J. (2012). *Unrelenting innovation: How to create a culture for market dominance.* John Wiley & Sons.

The Office of the High Commissioner for Human Rights. (2024). *What are human rights?* United Nations. Retrieved August 14, 2024, from https://www.ohchr.org/en/what-are-human-rights

The World Bank. (2022, June 27). *Stagflation risk rises amid sharp slowdown in growth* https://www.worldbank.org/en/news/press-release/2022/06/07/stagflation-risk-rises-amid-sharp-slowdown-in-growth-energy-markets

Thukral, N., & Christina, B. (2022). Analysis: As wheat prices soar, the world's consumers vote with their feet. *Reuters—Commodities.* https://www.reuters.com/markets/commodities/wheat-prices-soar-worlds-consumers-vote-with-their-feet-2022-08-03/

Unglesbee, B. (2024). Shortages 2024: What supplies are still at risk after years of disruption? *Supply Chain Dive.* Retrieved November 17, 2024, from https://www.supplychaindive.com/news/shortages-2024-supply-chain-risks-food-medical-supplies-minerals-climate-change-geopolitics/704051/

United Nations Inter-Agency and Expert Group on SDG Indicators (IAEG-SDGs). (2024). *The sustainable development goals report 2024.* U. N. Publications. https://unstats.un.org/sdgs/report/2024/The-Sustainable-Development-Goals-Report-2024.pdf

United Nations Office for the Coordination of Humanitarian Affairs. (2021). *Inside Yemen.* United Nations Office for the Coordination of Humanitarian Affairs. Retrieved August 19, 2022, from https://insideyemen2021.unocha.org/index.html

van Waterschoot, W., & Van den Bulte, C. (1992). The 4P classification of the marketing mix revisited. *Journal of Marketing, 56*(4), 83–93. https://doi.org/10.1177/002224299205600407

Vargo, S. L., & Lusch, R. F. (2008). Service-dominant logic: Continuing the evolution. *Journal of the Academy of Marketing Science, 36*(1), 1–10. https://doi.org/10.1007/s11747-007-0069-6

Vargo, S. L., & Lusch, R. F. (2018). *The Sage handbook of service-dominant logic.* Sage.

Veblen, T. (1998[1899]). *The theory of the leisure class.* An Economic Study of Institutions. Prometheus Books.

Wells, H. G. (2016[1914]). *The war that will end war*. Read Books.

Wieseke, J., Geigenmüller, A., & Kraus, F. (2012). On the role of empathy in customer-employee interactions. *Journal of Service Research, 15*(3), 316–331. https://doi.org/10.1177/1094670512439743

Wiliński, J. (2017). War metaphors in business: A metaphostructional analysis. *ANGLICA-an International Journal of English Studies, 26*(2), 61–78.

Woodall, T. (2012). Driven to excess? Linking calling, character and the (mis) behaviour of marketers. *Marketing Theory, 12*(2), 173–191.

Wright, R. (2024). Russian tourists are banned in most of Europe, but this country is welcoming them with open arms. *Euronews.* https://www.euronews.com/travel/2024/02/09/russian-tourists-are-banned-in-most-of-europe-but-this-country-is-welcoming-them-with-open

Wunder, T. (2023). *Essentials of strategic management: Effective formulation and execution of strategy in the era of sustainability.* Schäffer-Poeschel.

Young, C. (2024). Wheat shortage? Use more of it, says. *Real Bread Campaign.* Retrieved November 18, 2024, from https://www.sustainweb.org/realbread/ourwork/

Zagout, A. (2023, September 9). *Poverty in the State of Palestine set to soar by more than a third if war continues for a second month.* United Nations Development Programme. Retrieved November 18, 2024, from https://www.undp.org/press-releases/poverty-state-palestine-set-soar-more-third-if-war-continues-second-month

Zarkada-Fraser, A. K. (2001). Stereotyping in international business. In C. Cooper, S. Cartwright, & C. Earley (Eds.), *The International handbook of organizational culture and climate* (pp. 391–406). John Wiley and Sons.

Zarkada-Fraser, A. K., & Fraser, C. (2002). An exploratory investigation into cultural awareness and approach to negotiation of Greek, Russian and British managers. *European Business Review, 14*(2), 111–127. https://doi.org/10.1108/09555340210420091

Zarkada, A. K. (2013, September 23–25). Adapting to Survive: Facebook's introduction into the IMC Ecosystem 4th EMAC CEE Regional Conference, St Petersburg.

Zarkada, A. K. (2021, June 24–26, 2021). *Listening to a unicorn's pink narrative: Scheduling of brand Posts on Instagram for maximum consumer brand engagement* 19th International Conference on Research in Advertising (ICORIA) virtual conference.

Zarkada, A. K. (2024). Marketing solutions to the challenges of a VUCA environment: What have we accomplished and what more can we do? In A. K. Zarkada (Ed.), *Marketing solutions to the challenges of a VUCA environment.* Springer. https://doi.org/10.1007/978-3-031-58429-9_1

Zou, S. (2015). Political risk assessment: A review and a new approach. In M. Levy & D. Grewal (Eds.), *Developments in marketing science: Proceedings of the academy of marketing science* (pp. 245–247). Springer International Publishing. https://doi.org/10.1007/978-3-319-13159-7_53

Ευρωπαϊκή Επιτροπή (European Commission). (2012, July 25). *Σκάνδαλο Libor: Η Επιτροπή προτείνει δράση σε επίπεδο ΕΕ για την καταπολέμηση της χειραγώγησης των επιτοκίων* https://ec.europa.eu/commission/presscorner/detail/el/IP_12_846

Πολυχρονιάδης (Polychroniadis), M. B. (2019). 20 χρόνια από τη χρηματιστηριακή «φούσκα» του 1999 (20 years from the stock exchange "bubble" of 1999). *LIFO.* https://www.lifo.gr/now/greece/20-hronia-apo-ti-hrimatistiriaki-foyska-toy-1999

Ο Περί των Συμβάσεων Πίστωσης για Καταναλωτές σε σχέση με Ακίνητα που Προορίζονται για Κατοικία Νόμος του 2017 (N.41(I)/2017), (2021). http://www.consumer.gov.cy/meci/cyco/cyconsumer.nsf/All/2D72C3F501566901C22586BD0041E95E/$file/%CE%91%CF%80%CF%8C%CF%86%CE%B1%CF%83%CE%B7%202021-01%20(%CE%95%CE%A0).pdf?OpenElement

Χαροκόπος (Harokopos), K. K. (2021). Το Κραχ του '99 μας στοιχειώνει ακόμα (The '99 Crash is still haunting us). *Liberal.* Retrieved September 28, 2022, from https://web.archive.org/web/20210920223256/https://www.liberal.gr/market/to-krach-tou-99-mas-stoicheionei-akoma/402786

Out of the Permacrisis and Towards Sustainable Prosperity

2

Acronyms

MNCs	Multi-National Corporations
SMEs	Small and Medium size Enterprises

We visualise each and every living being as a twig, on the branch of a tree, in a dense forest. Some might be leafy, some sparse, or dry, some high up, and others low hanging, but we all belong to species, races, families or herds, countries, cities, communities, professions, religions, or organisations—the trees of the forest and the branches of the trees which are our inescapable ties with the groups we choose or happen to belong to. If the tree is on fire, all twigs will burn too. If the twig catches fire, it can burn the whole forest down to ashes and no twig can stay green in a burning forest. The Los Angeles fires which ushered in 2025 for hundreds of thousands of people whose houses were burned to the ground painfully proved that no gates, alarms, cameras, and armies of private security firms can save even the richest and most famous from nature, climate change, and public spending cuts (FitzGerald & McArthur, 2025) (Fig. 2.1).

Not all is doom and gloom, however. We know very well what we are doing wrong, and this means that we can change it. Just like rain falls on the whole forest and keeps it alive and thriving, mindful action taking all stakeholders into account generates sustainable prosperity for all life and keeps it and the planet thriving. Floods, megafires, pandemics, and refugee crises amply demonstrate that the fate of each and every person is dependent on the health, safety, and prosperity of everybody and everything else on the planet. Individual health adds up to public health, but also it only takes one person getting sick in a remote province of China to kill over seven million people all over the world (World Health Organisation, 2025).

Just as the absence of disease is not enough to be healthy, the absence of poverty is not enough to be prosperous. Just like physical well-being, financial security and

A. K. Zarkada and Y. Melanthiou, *Future Management for Sustainable Prosperity*, Future of Business and Finance, https://doi.org/10.1007/978-3-032-15868-0_2

Fig. 2.1 Visualisation of global interdependence. Computer artwork by Anna K. Zarkada, 2023

prosperity are both an individual and a social matter, and the two are inextricably intertwined. At the collective level, prosperity presupposes that production and consumption are, at least, sustainable. Their being sustainable means that they serve the needs of the people at the current point in time without compromising the production and consumption potential of future generations. It means taking from the physical environment and from each other only what is needed, redressing the damages done by previous generations, and ensuring the resources are replenished at a rate capable of ensuring the sustenance and prosperity of the generations to come.

Sustainable development is realised through the careful management at the local level of resources that are, however, recognised as universal and common to all, including future generations. It also presupposes that we develop while recognising the rights of developing countries to rise out of poverty. As Norway's first female Prime Minister and former Director-General of the World Health Organization, Gro Harlem Brundtland noted, well-being means a quality of life achieved through education, employment, social justice, and social security with equality on a global basis and global solidarity in mind (Brundtland, 1994).

2.1 The Concept of Prosperity Through the Ages and Traditions

Prosperity, also denoted with terms such as happiness, well-being, living the good life, bliss, or terms borrowed from other languages such as eudemonia (Greek), hygge (Danish), or ikigai (Japanese), has been examined by philosophers, religious leaders and theologians, social scientists, psychologists, and artists for thousands of years. We choose the term prosperity, which comes from the Latin *prosperāre*, meaning to be happy, fulfilled, and flourishing. It is the closest to *εὐημερία* (*εὖ* = good + *ἡμέρα* = day), the word used in ancient Greek philosophy for living a comfortable, meaningful, fulfilled, and good life. Prosperity encompasses how one spends one's life, achieving one's purpose, and leaving this world in a shape better than the shape it was in when one came to it. Prosperity thus covers a person's physical life and extends to how one is remembered after one's physical life is completed, what is known as *ὑστεροφημία* (posthumous fame).

Croesus, the king of Lydia whose name went down in history as a synonym for unbelievable and ever-increasing lands, tithes, and material wealth, boasted that he was the happiest man in the world. When the philosopher Solon was asked to name the happiest man on earth, he refused to name Croesus. Croesus challenged him and Solon replied with the Delphic dictum «*μηδένα πρὸ τοῦ τέλους μακάριζε*» (naught before the end rejoiced). Indeed, the end of Croesus proved the philosopher right. In preparation for an attack on the Persians, Croesus consulted the oracle in Delphi. He misinterpreted Pythia's utterance «*Ἅλυν διαβὰς μεγάλην ἀρχὴν καταλύσει*» (if you cross the river Halys a great power will be subdued), attacked, and lost the battle and his whole kingdom (Ηρόδοτος (Heredotus), 1836/[450 BCE]). Thus, it was proven that wealth and conquests do not equal prosperity, neither do they guarantee the sustainability of even a subjective feeling of happiness.

Solon also gave an example of what he believed to be the happiest man ever. He identified the Athenian king Tellus, who had good and worthy sons, lived to see his children's children born, all strong and healthy, and his life was sealed by a glorious death: he fought and died like a soldier, and the Athenians buried him with great honours and in full glory (Ηρόδοτος (Heredotus), 1836/[450 BCE]). Thus, the Athenian statesman and poet positions material wealth as secondary to moral integrity, reputation, and an enduring legacy in the pursuit of prosperity.

Aristotle, in the *Nicomachean Ethics* (Aristotle, *trans.* 1954/[fourth century BCE]), described prosperity as

- the ultimate goal and ultimate objective of all human action,
- a general state of satisfaction, and
- success,

achieved through the exercise of the practical virtues, namely:

- prudence,
- courage,
- moderation, and
- justice (Scruton, 2012).

In the *Rhetoric* (Aristotle, *trans.* 2004/[fourth century BCE], pp. Part V, 10), Aristotle emphasised that we should do what creates or increases prosperity and avoid what destroys it or prevents it from developing. He stated that "pretty well everybody agrees" that prosperity "is one or more of"

- independence of life;
- the secure enjoyment of the maximum of pleasure
- a good condition of property and body,
- the power of guarding one's property and body and making use of them.

He also listed "the constituent parts of prosperity":

- good birth,
- plenty of friends, good friends,
- wealth,
- good children, plenty of children,
- a happy old age, also
- such bodily excellences as
  - health,
  - beauty,
  - strength,
  - large stature,
  - athletic powers, together with
- fame,
- honour, good luck, and
- virtue.

He stressed the need for balance of internal goods of soul and body, such as beauty and virtue, and external goods such as friends and honours. Finally, he believed that to experience prosperity "one must have resources and luck in order to make one's life truly secure".

The Latin Stoic philosopher Seneca devoted his book *De Vita Beata* to prosperity, the good life, which he emphatically distinguished from the mere accumulation of wealth. His phrase "*Hominimum bonum quaero, non ventris*" (I seek the good of man and not the good of his belly) has become proverbial (Seneca, 50 BCE / 2004, pp. 56–57). He regarded philosophy as a balm for the wounds caused by the destructive passions, especially anger and sorrow. He saw the good life as the balance between an active life and contemplation. He also believed that prosperity is a result of the disassociation from the pursuit of wealth and dependent on the proper use of one's resources for the benefit of other people and society. Finally, he linked the achievement and maintenance of prosperity—what we would call today sustainable prosperity—to treating people with leniency, showing friendship and solidarity, and practicing moderation (Colish, 1990).

Modern philosophers also linked prosperity with virtue. James Fitzjames Stephen, for example, expressed the view that prosperity consists of knowing that we belong to a secure moral order which safeguards our right to expect that our fellow citizens will always act according to this moral order (Scruton, 2012). Bertrand Russel (Russell, 1930) viewed prosperity as a result of achievement of success at something which seems hard to achieve. He also saw playing a role in communal life and taking part in the writing of history or the creation of a new world as the source of the profound satisfaction required for prosperity. As he aptly put it: "[t]he happiness of the reformer or revolutionary depends upon the course of public affairs, but probably even while he is being executed he enjoys more real happiness than is possible for the comfortable cynic". To summarise Russel's key points, prosperity results from:

- success in one's efforts
- belief in a cause,
- a friendly interest in persons and things,
- companionship and cooperation,
- affection which "is reciprocally life-giving; each receives affection with joy and gives it without effort, and each finds the whole world more interesting in consequence of the existence of this reciprocal happiness",
- living objectively, and
- taking part in communal pursuits
- genuine zest for "passions and interests ... directed outward, not inward" (Russell, 1930).

These observations are also consistent with the views of contemporary spiritual leaders, such as His Holiness the Dalai Lama. He positions being happy as the purpose of life and sees the source of all happiness in compassion and warm-hearted kindness towards others which bring us peace of mind (His Holiness the 14th Dalai Lama, 2021). Similarly, Pope Francis, in his homilies, equates selfishness, the "spend[ing] one's own talents, one's energy and one's time only to save, protect and fulfil oneself" with "losing oneself, i.e. to a sad and barren existence (Pope Francis, 2017)".

Even contemporary Ecological Economists summarise their analyses with similar observations:

> The truth is, a genuine prosperity transcends material concerns. It consists as much in our social and psychological well-being as it does in our access to material things. It rests on the love of our families. It is inherent in the respect of our friends. It depends on the resilience of our communities. It resides in our ability to live a life without shame, to participate fully

> in the life of society. And though the restless marketers of consumerism persuade us endlessly that these vital goals can only be met with their own indispensable products, social psychology itself defies this pathological logic. (Jackson, 2015, p. xiii)

Ordinary people also seem to share the views of the great thinkers in increasing numbers. In a 2014 survey of how US citizens perceive the 'American Dream', only 23% associated it with material wealth. Most considered it to be the satisfaction of their basic needs (75%), making use of the full range of their potential (71%) and being in harmony with nature (54%) (Urban Sustainability Directors Network, 2014). In 2024, young Americans were found to "prioritize happiness over just money, value family, … aim for a stable future [and generally present as] a resolute generation, hungry for balance and fulfillment" (Pitts, 2024).

Balance and fulfilment are at the core of the concept of *Ikigai* (生き甲斐 - 生き (pronounced iki) which means life and 甲斐 (kai) which means worth), life's worth. The concept originated from the Japanese islands of Okinawa, which are famous for having one of the highest concentrations of centenarians and some of the lower rates of chronic illnesses and dementia in the world. It has found its way into Western explorations of prosperity as one's reason for being, "the reason you get out of bed in the morning" (Tamashiro, 2019). Ikigai is the intersection of what you love and are good at with what the world needs and will gladly reward you for. Passionately and skilfully pursuing the provision of value and finding fulfilment in service is a life of meaning which results in prosperity, of both the person who chooses to live a life of purpose, and of the community the person serves.

2.1.1 Dimensions of Prosperity

Prosperity manifests in three distinct but interrelated dimensions: health, thriving, and wellness.

The World Health Organization defines health, one of the fundamental human rights, as «a state of complete physical, mental and social well-being and not merely the absence of disease or infirmity» (World Health Organisation, 2022). As a human right, health is a basic good that all human beings should enjoy without discrimination on the basis of race, religion, political opinion, economic or social status, age, gender, or any other characteristic. Health depends on the fullest possible cooperation of individuals and states. Governments have a responsibility to provide for the health of their people, a responsibility which can only be fulfilled by providing adequate resources to support health and social welfare institutions which are open and accessible to all people within their jurisdiction. The public also has a responsibility to actively cooperate with the authorities after forming an informed opinion based on valid and timely information provided by the relevant agencies (World Health Organisation, 2022). The concept of health encompasses good physical, mental, and social health and the efficient and effective management of the internal and external factors affecting them. Central to achieving and maintaining mental and social, but also physical health is the individual's ability to communicate with others in order to cooperate, resolve disputes, and build and maintain interpersonal relationships and social bonds based on respect and affection.

Thriving is the achievement and maintenance of security in terms of access to the resources necessary for maintaining health, comfort, personal growth, and the engagement with people or things one is interested in, as well as having enough to contribute to one's family, community, and society in general. Finally, thriving is fundamentally temporal; it manifests in meeting present goals, managing risks and future threats, being able to deal with unforeseen adverse circumstances, and providing for future needs. It is also closely associated with wellness.

Wellness is the subjective experience of feeling healthy and comfortable in one's body, within one's mind, and with the world. Wellness requires being able to effectively manage life's problems and adversities. Finally, it has to do with enjoying one's place in the lives of others, feeling and returning compassion, affection, appreciation, and respect, and experiencing the security that these feelings are not fleeting but solidly based on fitting comfortably within society, and being in harmony with the world.

2.1.2 Prosperity Is Inherently Universal

In all the definitions and conceptualisations of prosperity we discussed above, other people and groups of people—living, dead, and not born yet, in one's immediate environment or far away and never to be acquainted with—have a prominent place. The prosperity of the individual is thus a result of the proper functioning of the person's relationships with other people and groups as well as a contributor to, but also a result of, collective prosperity. Like the twig in the forest (Fig. 2.1), one cannot prosper if other people suffer. We experienced it first-hand with the COVID-19 pandemic. Even the people who did not get infected suffered the practical and psychological consequences of mandatory isolation and restrictions on travel. They also experienced intense and prolonged fear and worry about their health and that of their loved ones. Moreover, no matter how much money they had, they could not enjoy it as they normally would because of the limited services available and the serious product shortages in the market.

What the world is experiencing during the permacrisis has been described by Bertrand Russel:

> All unhappiness depends upon some kind of disintegration or lack of integration; there is disintegration within the self through lack of coordination between the conscious and the unconscious mind; there is lack of integration between the self and society where the two are not knit together by the force of objective interests and affections.

He continues with the positioning of prosperity as an inherently universal condition, in the final words of *The Conquest of Happiness:*

> The happy man is the man who does not suffer from either of these failures of unity, whose personality is neither divided against itself nor pitted against the world. Such a man feels himself a citizen of the universe, enjoying freely the spectacle that it offers and the joys that it affords, untroubled by the thought of death because he feels himself not really separate from those who will come after him. It is in such profound instinctive union with the stream of life that the greatest joy is to be found. (Russell, 1930)

The "union with the stream of life (Russell, 1930)" is a seamless one, with life constantly flowing from the individual to society and the world and back to the individual. No one has expressed this better than Nikos Kazantzakis:

> *"Όλοι είμαστε ένα, όλοι είμαστε μια κιντυνεύουσα ουσία. Μια ψυχή στην άκρα του κόσμου που ξεπέφτει, συντραβάει στον ξεπεσμό της και την ψυχή μας. Ένα μυαλό στην άκρα του κόσμου που βυθίζεται στην ηλιθιότητα, γιομώνει τα μελίγγια μας σκοτάδι."* (Καζαντζάκης (Kazantzakis), 1985/[1927], p. 88)
>
> *(We are all one, we are all an imperilled essence. If at the far end of the world a spirit degenerates, it drags down our spirit into its own degradation. If one mind at the end of the world sinks into idiocy, our own temples overbrim with darkness* (Kazantzakis, 2012/[1927]))*.*

It is not only with other people, however, that prosperity needs to be shared for it to be truly sustainable. Since the time of the Neo-Malthusians, it has been empirically established that nature (people, land, atmosphere, seas, lakes and rivers, and animals and plants) is not an agglomeration of disparate and amorphous 'resources' to be used as needed for the attainment of economic goals. It is a fully integrated complex system of multiple interdependencies. Any intervention in one part of the ecosystem brings about changes in a multitude of other interconnected parts. A prime example, of these interdependencies, is the story of *Australia's war against rabbits* (Cooke, 2014).

A Case in Point: You Cannot Win a War Against Yourself

In 1788 the Europeans and Africans who arrived in Australia with the First Fleet (the first 11 ships which brought 1,400 convicts, civil servants and settlers) brought with them the first rabbits to ever set foot on the continent. Rabbits were transported in cages to provide fresh meat during the months-long journey and, in the early years, were widely bred in yard or farm enclosures as food for humans and their principal partners in developing the primary industries, the working dogs. Moreover, they were an important source of additional income for the first settlers as their skins were used to make the iconic Akubra hat. The Akubra was, and still is, the essential protective gear for farm workers, a symbol of "the spirit of adventure … inviting wearers to be part of a legacy that celebrates the Australian identity and the adventures that define it" (Akubra, 2024), and, thanks to the success of *Crocodile Dundee*, a major Australian export for the affluent country chic around the world and a beloved souvenir for tourists to Australia.

Rabbits bred like rabbits and happily coexisted with people and indigenous flora and fauna, restrained in their enclosures, until about the middle of the nineteenth century when a bored and homesick settler released some rabbits from their enclosures to use as game for hunting. Hunting was a popular pastime of the time, so other settlers followed suit. Hunters and local carnivores managed to keep the wild rabbit population under control for about a decade. Rabbits, however, were quick to adapt to their new environment and became hardier. At the same time, the number of their natural predators decreased because people hunted the wild carnivores to protect their flocks of sheep. The result was the fastest spread of a mammal in recorded history.

The negative effects of humans introducing a new species into an ecosystem soon began to show. To feed themselves, the wild rabbits first exhausted the natural vegetation of the pastures, thus depriving the sheep and cattle of food. Then, they ate shrubs, newly planted trees, and all the leaves and bark of old trees they could reach, thus destroying orchards and disrupting the natural renewal of forests and the bush. Eventually, by eating the native plants, which left the soil exposed to rain, streams, and wind, they caused serious soil erosion problems.

Australian state governments from the 1880s onwards offered large rewards to anyone who could find a way to exterminate the rabbits, and by 1901 the problem had grown to such proportions that a Royal Commission was set up to assess the situation. Various methods were tried: deep ploughing of the lands where the rabbits lived, poisoning their burrows, and keeping them off the lakes from which they drank water with fences, which, although often hundreds of kilometres long, did not work as the rabbits would dig under them to get to the water. The reward for bringing in dead rabbit feet was also tried, but, no matter how many were killed, the numbers did not decrease. The trade in rabbit extermination products flourished for decades, supported by extensive advertising, mainly in newspapers, but also with street and shop window posters. Some typical examples of print ads for rabbit extermination products have been preserved in the digitised archives of the National Library of Australia and are presented in a collage here (Fig. 2.2).

Nothing worked, so they released cats, which also arrived in Australia with the Europeans in 1778 and until then lived in homes as pets and mouse hunters. People were hoping their cats would be more effective hunters than themselves. The cats, however, were not interested in hunting rabbits, which were too large and too fast for them. Instead, they hunted smaller animals, birds, and reptiles. Finally, the government and local authorities experimented with various viruses and, in 1950, succeeded to reduce the number of wild rabbits from 600,000,000 to 100,000,000 using the myxoma virus brought from America. Myxomatosis, of course, also killed many domestic and pet rabbits. Moreover, the small victory achieved after more than 70 years of war on rabbits was not permanent. It took rabbits about 40 years to adapt to the virus and become resistant to the disease, and so, wild rabbit numbers rose again to 300,000,000 by the end of the twentieth century (Cooke, 2014).

The war is still ongoing, but its painfully obvious collateral damages are dramatic and not at all restricted to Australia. The impact of rabbits on Australia's

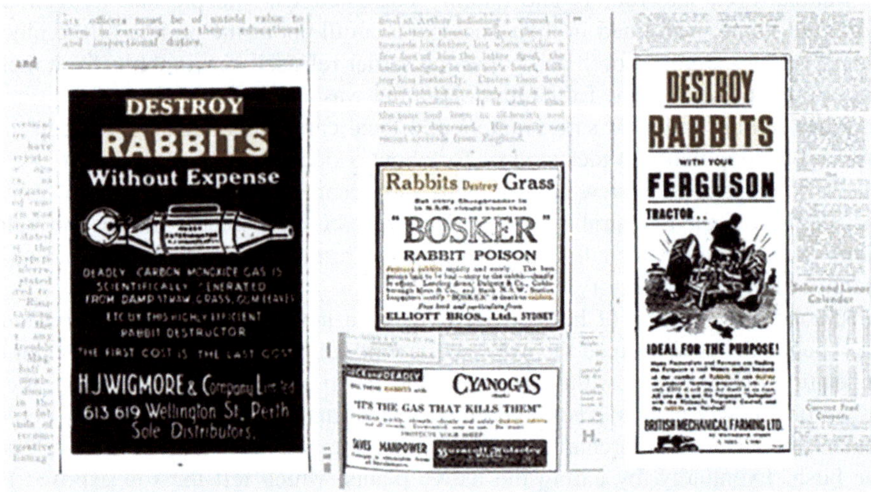

Fig. 2.2 Advertisements of rabbit extermination from early twentieth century. Source of the photos: the digital archives of the National Library of Australia (www.trove.nla.gov.au, free in the public domain). Computer generated collage by Anna K. Zarkada.

ecosystem has been, and remains, simply devastating. Some experts argue that feral rabbits are the most significant known cause of destruction of Australia's native flora due to overgrazing. Even if all rabbits disappeared tomorrow morning, it is estimated that it will take many hundreds of years for the soil to regenerate and reverse the desertification of vast areas destroyed by grazing rabbits. The effect on fauna has also been devastating. Cats, especially the feral cats living outside the urban fabric, kill 1,067,000,000 mammals, 399,000,000 birds, 609,000,000 reptiles, and 92,000,000 frogs every year. They are also responsible for the total extinction of many native species and at least 22 endemic mammals of Australia, such as the bilbies. No one seriously threatens the cats, who are patient, quiet, and versatile hunters, against which the native small species are simply unable to defend themselves. Finally, the myxomatosis and haemorrhagic disease of rabbits, which have now spread throughout Australia, in addition to domestic rabbits in Australia, have also infected rabbits on other continents. Their spread in Europe, for example, threatens with extinction protected species, such as the Iberian lynx and the Iberian eagle, whose survival depends on the abundance of healthy small mammals such as rabbits and other rodents, which, however, are being decimated by the introduced diseases (Invasive Species Council, 2022).

If the early settlers to Australia could claim they had no way to estimate the consequences of their actions—from introducing alien species to an ecosystem, to releasing domestic animals and deadly viruses to reverse the damages they had caused—the people of the twenty-first century can make no valid excuses for the negative externalities of their actions. There is enough recorded history, as well as ample data and the knowledge and machines to utilise them, for accurate forecasts of the consequences of our actions or omissions to be made. It follows, that it is the responsibility of businesses, and the governments who are supposed to be regulating their operations, to calculate the long-term consequences of any new product or process on a global scale. This is what algorithms, AI, and digital twins are for, anyway.

2.1.3 Prosperity Is Inherently Sustainable

The human species, despite its huge footprint on the planet, is only one of the estimated 8,700,000 species of plants and animals in existence (National Geographic Society, 2022). It stands to reason that, just as it is not possible to save only a twig when a forest is burning, it is not possible for a species to prosper when others are becoming extinct, neither can a single human being prosper in a suffering society. Such a short-sighted view is at the root of the ills of modernity from which the world will not be relieved unless we change our perspective and our practices. To achieve this, we need to regenerate our mindsets towards our stakeholders.

Sustainability is inherent in prosperity as we can only prosper when we act in ways that ensure the health, thriving, and wellness of not only those existing in the present, but also of those coming in the future. It is our responsibility to safeguard the right of all the coming generations of humans and other species to inherit a viable natural environment and an economy which is not burdened by the debts we have created. We need to redefine who the stakeholders of each of our decisions and actions are and consider many more than the immediately obvious. It is not enough to stop doing harm; we are also responsible for reversing the damages we have caused and also assume responsibility for those caused by previous generations. To achieve this, we need to regenerate our mindsets towards time.

Prosperity is inherently sustainable, a state of being without time boundaries, because societies exist beyond the limits of the physical lives of each of their members and ecosystems existed before each one of the organisms in them was born and will continue to exist after each one dies. Each social group and the sum of the social groups that make up our world come from the depths of history; they bring with them traditions and customs, but also stories and memories much older than the memory of the oldest living person, and will continue to exist in the future, long after the death of the youngest person now living on earth. Methuselah, a Great Basin bristlecone pine tree in California, is believed to be almost 5,000 years old, "meaning it was well established by the time ancient Egyptians built the pyramids at Giza (Bauza, 2023)". We are as responsible for its prosperity as we are for our lawn and our children.

Prosperity is by definition universal and sustainable. It starts with a change of mindset of one individual at a time, and spreads to the family, the work environment, the wider community, and ultimately the world. The mindset of respect for those who have gone before us and concern for those who come after us is the foundation of sustainable prosperity for all.

2.1.4 Sustainable Prosperity: The Goal of All Economic Activity

The term sustainable prosperity has been used by economists to describe a globally balanced economic development, one which provides stable employment for all and a fair distribution of income (Lazonick, 2009). The problem of fair distribution has been solved by the Apostles (Ελληνική Βιβλική Εταιρία, 2003), early Christian communities, nineteenth-century socialists (Bovens, 2020), Marx (Bovens, 2020), and the constitutions of communist countries (Τόλιος (Tolios), 2021): from each according to ability; to each according to need.

A globally balanced sustainable economy geared towards prosperity develops like a living organism: it evolves and grows better, stronger, and fairer. It neither just inflates like a balloon, nor do some parts of it expand at the expense of others. It gets better with time, which means it offers more space, time, resources, and opportunities for people to prosper. It fosters interpersonal human relationships and supports the development of intergroup collaboration. It provides efficient and transparent mechanisms capable of facilitating free, honest, and productive communication, encourages and supports value cocreation, and maintains procedures for equitable distribution of value and fair dispute resolution which are accessible to all.

Sustainable prosperity is not quantified through an aimless numbers game based on arbitrary measures equating societies with profit-making cost centres (Moore et al., 2021; Stiglitz, 2020). Its evolution is reflected in how it improves living conditions for all, how its products serve the users better, and how it demonstrates respect for human rights, the other inhabitants of the planet, the planet, and the universe (which we are already littering with space debris and to which we seem to have begun exporting our problems). Like all living things, a world geared towards sustainable prosperity evolves, changes, adapts, and allows space for new forms of socioeconomic organisation to develop and evolve. It provides resources, allows time, and affords the freedom to all agents to innovate, experiment, and test new systems and management methods. Eventually, its achievements are superseded, and its socioeconomic systems are replaced by the most efficient of the novel systems it has nurtured into becoming.

2.2 Sustainable Prosperity Is a Collective Responsibility and a Management Task

Given the present permacrisis conditions, it is up to us, now, to employ Future Management to right the wrongs and repair the damage caused to the natural and human capital by the distorted and unbalanced patterns of twentieth-century economic development. Further to redressing the past, we need to pave the way forward. We need not wait for enlightened leaders to kindle world-changing revolutions; there probably aren't any, anyway. We need to do it, all of us, each and every one of us. As Nicos Kazantzakis wrote in *Ασκητική - Salvatores Dei (Ascetic - The Saviors of God):*

> *Ν' αγαπάς την ευθύνη. Να λες: Εγώ, εγώ μονάχος μου έχω χρέος να σώσω τη γης. Αν δε σωθεί, εγώ φταίω.* (Καζαντζάκης (Kazantzakis), 1985/[1927], p. 32)
>
> *(Love responsibility. Say: It is my duty, and mine alone, to save the earth. If it is not saved, than I alone am too blame* (Kazantzakis, 2012/[1927])*).*

To fulfil our duty to save the earth—and, we would add, ourselves and our children—we need to develop a new business and economic activity paradigm. This requires a systematic analysis of the characteristic structures of the status quo, a thorough examination of the permacrisis and its dialectic, and an accurate assessment of our position and power to change the status quo and pull ourselves out of the permacrisis in a peaceful and productive way. We have already gone through the first two steps of the process in the previous chapter. In this chapter we have outlined the status we believe we should aim for and discuss some approaches to achieving it through mindset regeneration.

Our method for doing this is by adopting the critical perspective premised on historical materialism, using the five analytical tools or guiding principles for understanding and interpreting complex, evolving situations regarding socio-political events and their inherent contradictions, organisational decision-making, and the material realities shaping our roots, present, and future. Most importantly, these analytical tools can serve as criteria for interpreting political, economic, or social events, labour markets and prevailing conditions, and the underlying dynamics and power relations behind structural changes, wars, environmental crises, or technological advancements. The same tools can guide the evaluation of the proposed underlying ideas and the assessment of the efficacy of the practical steps we outline below. These tools are the following:

- Wholeness. Contradictory developments need to be addressed in their totality instead of selectively choosing to focus on cases which fit a given position, a generally accepted 'truth', or the dominant narrative (Engels, 1998, 2001/[1925]). In the management vernacular, we should avoid missing the forest for the trees (or, even worse, our twig).
- Dominant Trend. To understand the dynamics of developments so as to be able to influence them, it is important to identify the most important factor, the dominant trend, driving them (Engels, 1998, 2001/[1925]). Examining the dominant trend reveals the forces of resistance and transformation. This allows alternative possibilities to emerge and can guide the focus of our energy and resources.
- Historicity. Developments do not take place in a vacuum. All current events and future developments are part of an ongoing historical process. Thus, events can only be understood in the context of their historical trajectory and the material realities which shaped them (Marx & Engels, 1888/1996). So, understanding what led to the current situation is fundamental to assessing its parameters and predicting its course.

- Class Analysis. The status quo is the result of the struggle between the classes. The power balance between competing interests is expressed through the developments, hence their being inherently contradictory (Marx, 2018/[1867]). Examining the classes and their interests is fundamental to understanding both the historical trajectory and the dominant trend and its sources.
- Partisanship. There is no such thing as an objective observer or a neutral analyst. To understand developments, one needs to be able to evaluate what they mean for oneself, the position one defends, and the catalysts needed to change the course of developments (Lenin, 1999/[1902]). This criterion emphasises evaluating events based on one's political goals and the actions required to advance those goals.

In the next section we discuss the design parameters of a system which can pull us out of the permacrisis and lead us to sustainable prosperity for all. We highlight the areas where we believe our mindsets need to be regenerated towards and the forces that can bring about the changes. The aim is to ensure the people of the twenty-first century and the coming generations will enjoy sustainable prosperity.

2.3 An Agenda for a Better Future: Restoring Métron

Traditionally, a basic pursuit of all systems for organising people into societies, such as cultures, religions, or political ideologies, is the control of human behaviour so that the desires and urges of one do not encroach on the rights and comfort of another. The key concept is that people have their basic needs fulfilled but with temperance and within given boundaries. For example, it is okay to have sex, but it must be within the boundaries of marriage; you should neither commit adultery, not even covet your neighbour's wife. Similarly, you can have oxen to plough your field, but you can neither steal nor covet those of your neighbours. Buddha (The Editors, 2025) is very much in agreement with Moses on the basics of human relations in the five moral precepts or virtues (*pañca-sīla*), namely:

- Abstain from killing
- Abstain from stealing
- Abstain from sexual misconduct
- Abstain from wrong speech
- Abstain from the use of intoxicating substances that cause inattention

The Yogi Patañjali teaches along the same lines in the five restraints (*yamas*): *ahimsā* (abstinence from injury), *satya* (truthfulness), *asteya* (abstinence from theft), *brahmacarya* (sexual control in thought, emotion, and action), and *aparigraha* (noncovetousness) (Catoméris, 2012). We could go on and on, around the world, but the point is that prophets and secular leaders alike have staged myth-creating spectacles as perception management mechanisms for ensuring that people adhere to their rules which are nothing but common sense. All societies adopt sets of rules and develop norms which ensure that "we treat humanity, whether in ourselves or in others, as a means only but always as an end in itself" (Johnson & Cureton, 2024). The Kantian (Kant, 1959/[1785]) introduction of the concept of 'respect' for persons, "for whatever it is that is essential to our humanity", reflects the Judaeo-Christian tradition which stems from Genesis 1:26 (NIV) (*Holy*Bible, 2011):

> *Then God said, 'Let us make mankind in our image, in our likeness, so that they may rule over the fish in the sea and the birds in the sky, over the livestock and all the wild animals, and over all the creatures that move along the ground'.*

Respect for humanity as the basis of the Categorical Imperative, that we are to "act only in accordance with that maxim through which you can at the same time will that it become a universal law" (Kant, 1959/[1785]) can serve social organisation very well. However, humanity has not proved to be a benevolent, not even a selfishly rational, ruler of all creatures. Human rule over the natural environment, and now space, is causing environmental degradation of extreme magnitude and extent—as the story of the rabbits in Australia we discussed earlier demonstrated—and its limits are manifested in the ecological disasters which seem to increase in frequency and severity. We propose that respect should be extended to encompass the whole of creation, and the maxims that we will to become universal laws need to be re-examined before it is too late. We believe that we have taken the blessing from Genesis 1:28 (NIV) (*Holy*Bible, 2011), to "fill the earth and subdue it" and greedily turned it into a curse.

One could argue that greed is also at the root of some people ruling over other people. We use 'rule' to denote brute power (the Weberian *Macht*) which is grasped by force or in any way exceeds the boundaries of rational-legal, or even charismatic or traditional authority (*Herrschaft*), that is, the boundaries within which the exercise of power is willingly accepted as legitimate by those subjected to it (Weber, 1978/[1922]). To protect people and social groups from the greed of those more powerful than them, all human communities have devised systems for the distribution of freedom and authority collectively referred to as rights. Rights determine the range of actions which are permissible and which institutions are perceived as desirable and just, and by accepting them, a society endorses "a certain view of what may, must, and must not be done", who can be in a certain state and who cannot, who can have what and how one is to acquire it. Subsequently, the set of rights a society endorses determines the structure and content of its government, law, and morality (Wenar, 2023).

Rights are a popular topic, and hotly contested issue, for philosophers, political scientists, and modern activists. Without going into details about their categorisation, form, function, language, justification (i.e. their being status-based, instrumental, or contractual and justificatory), relationships between rights (e.g. conflicts between rights or one right supporting another), or conflicts with other concepts (e.g. freedom), we need to stress here that their doctrine and language have been the subject of extensive critiques, mostly from the communitarian and

Marxist perspectives (Mackie, 1990; Waldron, 1984; Wenar, 2023). The preoccupation of modernity with fundamental rights and the postmodern confusion between rights and whims cloud the issue that there is nothing universal in the perception of "Who is alleged to have a right, … What actions or states or objects the asserted right pertains to, … Why the rightholder (allegedly) has the right, … [and] How the asserted right can be affected by the rightholder's actions [e.g.] The inalienable right to life, the forfeitable right to liberty, and the waivable right that a promise be kept (Wenar, 2023)".

One of the most recent rights systems, which is supposed to unite 27 countries and roughly 450million people, *The Charter of Fundamental Rights of the European Union* (The European Parliament, 2012), is an example of an ideologically driven structure, one firmly embedded in libertarian capitalism (its opponents, for example, would reject Article 17—Right to property and would see the control of all means of production as the exclusive right of the state or the commune). It also serves as demonstration of the relativity of principles expressed as absolutes as it fails to materialise as much more than a wish list (cf. Article 11.—Freedom of expression and information, with the blocking or Russian media (Hsu & Myers, 2024) and the coverage of the Palestinian perspective in the European press (Lubbock, 2024)).

This discussion aims to introduce the problematisation over human behaviour, its regulation, and consequences. Does the right to life of "an essentially worthless fish called a smelt (Jones, 2025)" trump (pun very intended) the right to property of Californian homeowners? Is any creature really 'worthless'? Should we blame climate change or public funding cuts for the devastation caused by the 2025 wildfires in Los Angeles (Jones, 2025)? These, and all other questions of this kind regarding geopolitics, technological developments and their application, economic development, or consumption behaviours, fail to provide the clarity we need to guide decision-making. But decisions we need to make, and fast! We also think that there is an excess of emotion and a dearth of reliable information clouding the most important of current debates regarding business and public management. Finally, we think that there are valuable sources of information and inspiration in a variety of philosophical and cultural traditions, if only we dare to look outside the twentieth-century management textbook box. This we attempt to do in the rest of this chapter, hoping to help the regeneration of mindsets that will support Future Management.

2.3.1 Defining Métron

In philosophy, the Greek word *μέτρον* (pronounced métron) is often translated as 'mean', 'measure', or 'moderation', but it actually means much more than just an average, or a way of estimating the size of something, or a restriction. It is a way of viewing the world and finding one's place in it on a moral, cultural, practical, and aesthetic level.

Métron is first and foremost the exercise of virtue. In its moral sense it is defined as a *habitus* formed by humans realising their natural potential for doing good through exercising rationality in their decisions, consciously practicing restraint of passions and urges, and acting in accordance with nature (Bejczy, 2011). For Aristotle and major streams of western philosophy, métron represents the perfect balance between excess and deficiency. Thus, métron is a standard of virtue expressed as having, doing, and enjoying enough but not too much. In this sense, métron is similar to the Swedish concept of *Lagom*, meaning 'just the right amount' or 'neither too little, nor too much', an attitude placing value on balance, moderation, and simplicity in all aspects of life (Mahajan et al., 2023).

The concept of Lagom is culturally determined and also a determinant of culture. People in Sweden describe everything, from their health to the strength of their coffee, using this small but loaded word. "*Lagom är bäst* (lagom is best) they say, rolling their eyes" (Knowles-Dellner, 2017). Thus, it is central to "Sweden's national psyche and characterises everything from Sweden's political leanings and stance on gender equality to their aversion to anything too ostentatious. It is at the core of many typically Scandinavian ideals like fairness, consensus and equality (Knowles-Dellner, 2017)". It has been argued (Lubowiecki-Vikuk et al., 2021) that these Scandinavian ideals are fundamental to the Scandinavian countries' topping the list of happy countries year after year (Helliwell et al., 2019, 2023). So, similar to Lagom, in terms of social organisation and the underlying logic of allocation of rights and responsibilities, métron determines that each person contributes to the common good as much as and however they are capable of and receives from society whatever they need to prosper. Finally, as a moral stance, métron emphatically encompasses the exercise of voluntary temperance and actively avoiding any extremes, as the excess of even a virtue is actually a vice (Bejczy, 2011).

In practical terms, métron denotes scale, as in size and its measurement. Métron defines the stimuli that are intense enough to be perceptible by the human sensory organs but not so intense as to damage them. It implies that messages and their meanings, or tasks and procedures, are neither so simple that they fail to attract and hold attention, nor too complex to be understood or so difficult that they cause frustration. Inherent to this meaning of métron is relativity. Stadium floodlights make attending a football game at night possible, but they would be blinding in a small room. Overly loud music becomes as disruptive as music that is audible but too quiet to be clearly discernible. A cluttered advertisement with excessive text and images may confuse or repel the audience and result in the message being lost. Métron is also contextual. A skyscraper in New York is just a building of appropriate size and shape, but in a rural village, among cottages, it would feel jarring and totally out of place. In every human enterprise and every interaction between people and organisations, métron can guide us to calibrate experiences and messages to fall within human perception, understanding, and comfort and to respond proportionately to challenges. The principles of relativity and contextuality in finding the right size and complexity apply to organisations. For example, businesses that are too small do not have the capacity to adequately serve their markets, but companies that are too large in relation to the economy that hosts them can end up having so much power that they distort the balance of democratic representation of concerns in a country and undermine the fair and transparent legislative process. Thus, métron enhances functionality, ensures engagement, safeguards efficiency, and allows harmony to support sustainable prosperity.

Métron also implies regularly recurring patterns and proportion. Thus, beauty is inherent in métron and métron is necessary for humans and their creations to be perceived as beautiful. For example, the size of each and every body part in relation to the whole body needs to follow the canon of proportions. Many mathematicians, artists, and philosophers—such as Polyclitus, Lysippus, Euclid, Luca Pacioli, Leonardo da Vinci, Le Corbusier, and Salvador Dali, to name but a few whose efforts have shined through time—have invested time and effort on calculating and elaborating the canon of proportions, also known as the golden ratio, or the divine proportion ($\varphi = 1.618033988\ldots$) (Carlson, 2024). The closer to the canon of proportions the ratio of the sum of the various body parts in relation to each other and to the whole is, the more a person is perceived as beautiful. Leonardo da Vinci's study shown in Fig. 2.3 illustrates the ideal human body and the calculations for reaching its measurements. How tall or short, sinewy or curvaceous a body needs to be to be considered as beautiful has changed drastically over the ages. The aesthetically pleasing proportions of the parts to the whole, however, have remained constant for millennia (Eco, 2004). Métron also gives us the proportions of the height and base of a building that make it habitable and pleasing to the eye, and the tread and riser dimensions a staircase must have to be comfortable to climb. Finally, it is métron which makes the difference between music and noise, clamber and dance, prose and poetry… the list is endless.

So, in short, métron defines what is good and what we can perceive, understand, and manage. The idea that violations of métron have disastrous consequences is found in all mythologies, which means the problem of how much is too much has troubled humanity since before the beginning of history. The Tower of Babel and the Cataclysm (Noah's, Deucalion's and Pyrrha's, Gilgamesh's, to name but a few) are characteristic examples of the need for human activity and behaviour to respect métron and refrain from violating it. The development of the ability to discern that which is 'just right' out of the available alternatives is so fundamental to the survival of the human species that it is being introduced to generations upon generations of children at their most tender age all over the world through the fairy tale of *Goldilocks and the Three Bears* and its many variants (Tatar, 2002). The story is a cautionary tale about finding not only the practical and aesthetic métron (as in the temperature of the porridge, the comfort of the chair, and the size of the bed), but also about the perils of carelessly applying one's métron on alien contexts (a human into the animal kingdom), and the moral and physical hazard of violating the métron set by the boundaries (the house) of the rights of others.

We here argue that all the problems we have so far discussed are caused by violations of the human métron on a global scale for too long. It could be argued that climate change is nothing but the twenty-first-century Cataclysm. The financial system is most likely another Tower of Babel. So far, we have seen banks

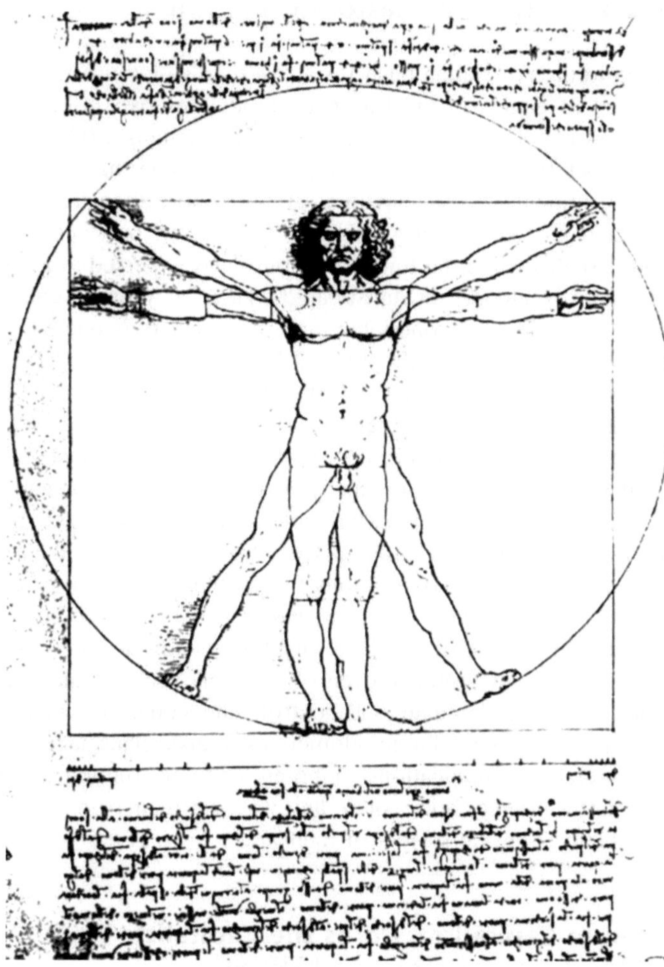

Fig. 2.3 Study of the proportions of the human body, known as the Vitruvian Man (c. 1509) by Leonardo da Vinci (Vinci 1452 - Amboise 1519) from the collection of prints of the Gallerie dell' Accademia Venezia Catalogue#: 228, Technique and material: Metalpoint, pen and ink, touches of watercolour on white paper. Available in digital form from https://www.gallerieaccademia.it/en/study-proportions-human-body-known-vitruvian-man. Public domain licence

which were too big to fail actually failing and taking with them whole national economies and countless livelihoods. The fact that taxpayers were forced to salvage small parts of the edifice does not mean that the edifice is still standing and growing taller. It also remains to be seen how the system will withstand the BRICS moving towards de-dollarisation and establishing another global payment system. In following sections, we offer insights from a variety of perspectives on how métron can be restored. We view these as general design parameters for the Future Management paradigm which can usher a time of sustainable prosperity for all.

2.4 Practical Steps for Restoring Métron

Rabbits released to the Australian bush for a few men's pleasure ended up threatening the survival of eagles in the European mountains over a century later. A small local act had major global consequences. As of August 2024, the US-listed tech giant, Alphabet Inc., the holding company of Google, has a market cap of $2.067 trillion (CompaniesMarketcap.com, 2024b), more than the GDP of Italy (Worldometer, 2024). Moreover, it grew by 31.05% in the past year (CompaniesMarketcap.com, 2024b) while the Italian economy was expected to grow by about 1% in 2024 (Istituto Nazionale di Statistica, 2024). After years in court, a federal judge ruled that "Google is a monopolist, and it has acted as one to maintain its monopoly". It "used [its] roots in the internet to influence the way we shop, consume information, and search online (McCabe, 2024)". It is likely that Google will need to be broken into smaller companies, which is what was supposed to have happened with Microsoft after it was also found to be a monopolist 20 years ago. Nevertheless, it still is the largest company in the world, with a market capitalisation of $3.157 trillion (CompaniesMarketcap.com, 2024a), larger than the UK economy and double than those of Australia and South Korea. So, the actual effect these rulings have on keeping corporations within the framework of competition or on providing consumers with access to better and cheaper products is dubious. It is obvious that born global private institutions can develop to a size way beyond what democratic institutions, even of the most socially and economically advanced nation-states, can regulate and control. The same, of course, is true for MNCs of all kinds such as media conglomerates, banks, pharmaceutical companies, agribusiness, telecommunications, shipping, and construction companies and their ability to lobby, coerce, bribe, or in any way get their own way. Rabbits or Google, the problem is the same: loss of métron.

2.4.1 When in Capitalism, Do as the Capitalists Were Meant to Do—Law and Ethics as Métron

Talking about corporations, markets, governments, capitalism, or the system, masks the fact that they are neither abstract constructs, nor shapeless forces like the wind. They are simply groups of people, so their acts are the acts of individuals; it is people with a face and a name and a career who make decisions. The NYT article *Inside VW's Campaign of Trickery* by Jack Ewing (Ewing, 2017) is full of names, the names of executives, engineers, secretaries, the people who decided to break the law, to lie, and to behave in specific ways. There is no such thing as markets; it is just traders and their clients who buy and sell shares, commodities, currencies, or repossessed houses.

In the contemporary economic system, there is no invisible hand; there are billions of hands that type the algorithms and press the buttons. And the people who control the hands have the power to do evil as much as they have the power to do good. MNCs and mom-and-pop stores alike are just vehicles for bringing the output of the labour of skilled individuals to other individuals. Managers and businesspeople serve the objectives of their organisations, for example, act so as to generate profits, "while conforming to the basic rules of the society, both those embodied in law and those embodied in ethical custom (Friedman, 1970)". None other than Milton Friedman in *Capitalism and Freedom* elaborates what he has called a "fundamentally subversive doctrine": "there is one and only one social responsibility of business–to use its resources and engage in activities designed to increase its profits so long as it stays within the rules of the game, which is to say, engages in open and free competition without deception or fraud" (Friedman, 1970, 1982). The métron in free market capitalism is imposed by laws (such as those regarding competition, arbitrage, collusion, bribery, tax avoidance and evasion, deceptive advertising, etc.) and 'ethical custom', and is restricted to the exchange of value through "voluntary cooperation with no coercion, deception or fraud and on the basis of the shared values and responsibilities of individuals who voluntarily form various groups that comprise society (Panigyrakis & Zarkada, 2014a)".

'If you think you are too small to make a difference, try sleeping with a mosquito' (Fig. 2.4). We do not know if it really is an old African proverb, or a Dalai Lama quote—he has, after all, on a number of occasions, discussed the ungratefulness and right to life of mosquitoes (Barry, 2016)—but I do believe that it is the cornerstone of the mindset of the Future Managers, the ones who will survive the Cataclysm. It might be true that governments cannot control the Googles of this world. As citizens we can vote for governments who will want to control them. We can also push for harsher laws and enforcement of existing legislation. But these might be long shots. The immediate action to take is to acknowledge that only their shareholders, employees, and customers can control businesses and act accordingly. No manager can break up monopolies singlehandedly, but every manager can refuse to participate in collusive or gouging practices (Ray et al., 1999; Zarkada-Fraser, 2000; Zarkada-Fraser et al., 1998).

2.4.2 Think Globally, Act Locally—Using Science and Technology to Determine Métron

The Scottish biologist, sociologist, and pioneer of town and regional planning Patrick Geddes, to whom the 'Think Globally, Act Locally' adage has been attributed, conceptualised 'eutopia' (literally meaning the good place) as "a place of effective health and well-being, even of glorious and in its way unprecedented beauty, renewing and rivalling the best achievements of the past, and all this beginning here, there, and everywhere" (Geddes, 1915, p. 73). Such a place would be the physical framework for sustainable prosperity for all. Be it a city, a country, or the whole planet, the eutopia "must be planned and realised, here or nowhere, by us as its citizens—each a citizen of both the actual and the ideal city seen increasingly as one (Geddes, 1915, p. vii)". The concept can be extended from ecosystems and cities to societies and teams, or economic systems and organisations.

Fig. 2.4 If you think you are too small to make a difference, try sleeping with a mosquito. Sketch by Anna K. Zarkada, 2025

Geddes proposed that the planning should be based on "the growing knowledge of Nature". The tools of science of his times (mid-nineteenth to early twentieth century) and the technologies at his disposal afforded far less effective and efficient means of calculating the present and future consequences of actions than contemporary ones. This, however, did not stop him from considering the global as well as local consequences of his proposals. In the time of super and quantum computing, AI, and Big Data, there is no excuse for seeking anything but "better use of resources and population" (Geddes, 1915, p. 73) by bridging technology with economics and sociology "to express in various ways the essential harmony of all these interests and aims; and to emphasise the possibilities of readier touch and fuller cooperation among them". Thinking of the global consequences of each and every action but keeping the decision making and actions within the locally manageable métron is the way in which ethics and economics, "these too long separated aspects of our conduct of life and of affairs may be reunited in constructive citizenship" (Geddes, 1915, p. v).

Citizenship is both the way to restore métron and the positive outcome of applying métron to the social, political, economic, industrial, and organisational structures we create or allow to develop. As citizens, we fulfil our purpose in life and contribute to communal welfare by:

- utilising our skills, talents, effort, and time as best we can and in whichever fields it is needed, that is, consciously and conscientiously practicing ikigai,
- acting according to the law, that is, respecting the social structure and staying within the métron of its legitimate authority,

- behaving according to generally accepted moral principles and social conventions, that is, respecting the ethical and cultural métron, and
- voluntarily participating in civic life and social causes.

We cooperate to achieve economic or social objectives, either informally or by forming associations and establishing institutions. These associations are also bound by the same obligations: to perform the function they were formed for (economic citizenship), to obey the law (legal citizenship), to act in socially acceptable ways (ethical citizenship), and to give something back to the society that hosts them (philanthropic or discretionary citizenship)—what came to be known as corporate citizenship (Carroll, 1998). Thus, métron determines the 'city' to which you can efficiently and comfortably be a citizen. The round the corner mom-and-pop store cannot do anything about potential violations of privacy and freedom of information rights by SpaceX, the owner of thousands, and soon to be tens of thousands, of satellites forming the Starlink constellation. Neither can they clean up the space debris and control the light pollution the overcrowding of low-Earth orbit (Wall Street Journal, 2022). Their 'city' is not the universe, not even their whole city; it is their neighbourhood. The mom-and-pop store owners, however, can respect the privacy and labour rights of their employees, they can dispose of packaging materials responsibly, and control the light pollution from their neon signs. They can also control the activities of the Space Xs of this world through their vote, and in this sense they can, and we here argue they have a responsibility to, think globally while acting locally.

Another way of expressing the concept of métron on a personal level is what Steven Covey, in his highly successful book *The 7 habits of highly effective people* describes as one's 'circle of concern' and 'circle of influence' (Covey, 2020). Our circle of concern includes all the issues we have some mental or emotional involvement with, i.e. "our health, our children, problems at work, the national debt, nuclear war". For some of them, like the national debt or nuclear war, we cannot do anything in the immediate future. We can, however, exercise daily, care for our children, and talk things over with our boss. All the things we are and can be, and the things that we can change through our actions comprise our circle influence. To be successful, we first focus on the things within our circle influence on which it is important to focus in the here and now. By systematically working on the elements of the circle of influence, and by being proactive, we can gradually expand our circle of influence to include elements which previously lay beyond our influence, in our 'circle of concern' (Covey, 2020). The more successful we are, the more power we have to expand the breadth and strength of our personal métron. Consistent local action gradually makes us 'citizens' of ever-widening locales, and increases our influence on the wider community, and eventually changes things on a global scale.

2.4.3 The Unbearable Permanence of Ephemeral Content—Phygital Communication Technologies in Need of Métron

In its simplest form, sustainable prosperity is the outcome of healthy, balanced, and symbiotic relationships between people. Its achievement depends on the relationships between individuals, between individuals and communities, and between communities. These relationships determine the division of activities and the distribution of the products these activities produce. Collective activity and the organisation of production and distribution create power structures and make up society. For centuries, these relationships were based on interpersonal communications, with mass communications being limited to official announcements and just whatever the town crier's voice and memory could carry. Even the violent imposition of the will of one group over that of another required physical contact as people fought with swords or rifle bayonets. Thus, our species became hard-wired to trust the senses and to believe our eyes and ears.

Inventions like print and radio introduced the technological intermediation to the communication process. Technology gradually increased the reach and decreased the personalisation of the message, but it was as late as the 1960s that mass media technologies established what we have come to know as the 'global village' (McLuhan & Powers, 1989). First, it was the news about major events, motion pictures, music, and gradually other cultural products such as television series, advertisements, and gossip about the lives of stars, presidents, and royalty that invaded the space of every household through the television set. In the twenty-first century, the lives of celebrities (irrespective of their being ascribed by virtue of bloodline, achieved through recognised talents and accomplishments, or attributed by attracting the attention of the media), as well as celetoids (who happen to be bestowed with a short but intense burst of fame because of good fortune or sheer bad luck) (Rojek, 2001), and even everyday people, also became mediatised spectacles to be admired, envied, ridiculed, despised, and discussed on social media.

Perceptions of Métron in the Digital Global Village

Nowadays, people's social lives—their friendships, degrees, weddings, and funerals—but also the most private aspects of their lives—the food and drink they consume, the brands of toiletries they prefer, the underwear they buy, and their illnesses—are exposed, literally, to the whole world and will remain there, accessible to those that control the platforms for ever. Further to the imagined communities of modernity (Anderson, 1991), metamodern people (Panigyrakis & Zarkada, 2014a, 2014b) belong to many different interest-, belief-, habit-, or memory-based digital communities, most of which are consumption related (Chen & Zhao, 2021; Flurry et al., 2014; Healy & McDonagh, 2013; Nightingale, 2020). Our ancestors were aware of details of the lives of their immediate family members, snippets of the lives of their personal acquaintances, and the most impressive fragments of the lives of their acquaintances' acquaintances. Their spatial boundaries were those of their village, their region, rarely their country, and only the very few (and very fortunate, or extremely unfortunate to have been embroiled in wars or to have had to emigrate to survive) were aware of faraway lands and their customs. We, on the other hand, have access to an unprecedented amount of vivid and detailed information about the physical and cultural world outside our immediate space through freely available audiovisual cultural products and relatively affordable travel. We also find ourselves connected to billions of people who watch the same Tik-Tok™ dances we do, we engage in political debates with them on X™, Like the same 5-min crafts videos on Facebook™, and shop from the same Instagram™ shops. Yet, it is, indeed, a small world we live in.

In 1929, the Hungarian author Frigyes Karinthy wrote in his short story *Láncszemek* (Chain-Links):

> *One of us suggested performing the following experiment to prove that the population of the Earth is closer together now than they have ever been before. We should select any person from the 1.5 billion inhabitants of the Earth – anyone, anywhere at all. He bet us that, using no more than* ***five*** *(sic) individuals, one of whom is a personal acquaintance, he could contact the selected individual using nothing except the network of personal acquaintances. For example, 'Look, you know Mr. X.Y., please ask him to contact his friend Mr. Q.Z., whom he knows, and so forth.'* (Karinthy, 1929)

The 'six degrees of separation' hypothesis was explored in the 1960s by Milgram (Milgram, 1967), and the Karinthy experiment was actually run by Travers and Milgram in the USA in 1969 (Travers & Milgram, 1969). It not only confirmed the 'small world' hypothesis, but, what is most important, it proved that people can skilfully navigate a complex social network they have no way of seeing in its entirety. Even if the number of steps between network members was later challenged, alongside the ability of 'friend of a friend' chains to span racial or class boundaries, it was acknowledged that, chances are, we will meet someone who knows someone we know. The more educated we are, the more likely it is to find a connection with any random person who travels in similar networks (Kleinfeld, 2002). Internet penetration rates renewed the interest in 'the small world problem'. A study of networks on a global scale using Facebook™ data and algorithms showed that the number of links needed to connect any two people was 5.28 hops in 2008, dropped to 4.74 in 2011, and could be as low as 3.0 degrees of separation when the analysis was limited to all connections being in the same country (Ugander et al., 2011).

The small-world feeling can be really comforting, even a literal lifesaver, especially in times of hardship such as the prolonged COVID-19 quarantines. At the same time, feeling part of a large network can be empowering. Most likely, the 721 m people who were active on Facebook™ in 2011 were not aware that they had formed 69bn friendships. Neither could they estimate that they could each reach an average of 156,569 other users at two degrees of separation, or that a typical, or median, user could reach over 31,000 people (Backstrom, 2011). They did, however, marvel at the number of Likes the videos of their cats got from long-lost classmates. They also, most likely, felt that they had the power to change the government when their political opinions were Shared by people they had never met. Back in the early 2010s, social media were, indeed, primarily social spaces. We met our old, and some really long-lost, friends and we befriended the friends of our friends. Some corporations were timidly making an entrance into Facebook™ but it was still in the fingers of the consumers to "allow companies into their private space if they prove[d] to be interesting, fun, and amazing enough to attract their Likes and to Share, that is, to market the companies' offerings to one another (Zarkada & Polydorou, 2014)". So, social media, despite their born-globalness and unprecedentedly staggering reach, were generally perceived as within the métron of human communities; they were media that mediated social interactions.

This is not at all the case in the 2020s. As of December 2024, 5.22 billion people, almost 65% of the total population of the world, spend, on average, over 2 h per day on almost seven different platforms (Singh, 2024). Most people focus on what they want to hear, the 'social', and fail to understand that social media are neither social, not really media; they are data collection and influence platforms owned by a tiny constellation of businesses. Each of these corporations is of a size well beyond métron in the sense that they are much more powerful than what most democracies can oversee and regulate. Moreover, they are becoming increasingly interconnected into an oligopoly which is not unlikely to end up merged into yet another tech monopoly that courts will try, and miserably fail, to break (Battelle, 2005; CompaniesMarketcap.com, 2024b; Fitzpatrick, 2014; McCabe, 2024). Because people choose to see social media as their phygital social space, they gladly contribute large amounts of current and fine-grained data in exchange for some attention, a little bit of company, or a short break from boredom. They choose to apply their own métron onto something way beyond any human métron with detrimental consequences. This was amply demonstrated in a recent scientific study:

> "(1) from 270,000 compromised accounts, Cambridge Analytica collected 68.0M Facebook profiles; (2) from surveilling 0.01% of the nodes in a mobile phone network, a law enforcement agency could observe 18.6% of all communications; and (3) an app installed on 1% of smartphones could monitor the location of half of the London population through close proximity tracing (Houssiau et al., 2023)".

2.4.4 Managing in the Phygital Domain

The findings of the Houssiau et al. study (Houssiau et al., 2023) are not the only evidence that understanding the mechanics and rationale of the digital domain is as important to management nowadays as is understanding the workings of the physical world, but much more complex, as it is fraught with the contradictions of total lack of transparency regarding the algorithms, little guidance from academia, and too much advice by consultants and self-appointed gurus. Thus, the managers of the twenty-first century, especially the digital migrants, stand before social media (SM), internet search engines (ISE), and artificial intelligence (AI) just as the medieval farmers stood before rain, hail, and locusts. They face significant opportunities as well as serious challenges in the digital and phygital domain at different personal and organisational levels. Some of the most critical tasks they have to perform are the following:

- Manage their personal image and influence (Littler, 2018; Zara, 2009).
- Design and implement policies to manage employees' personal and professional social media presence, including the control of their inputs in brand-related conversations with the public (Fuduric & Mandelli, 2014).
- Balance the gains in job satisfaction and creativity with the negative consequences, such as fatigue and role ambiguity, of Enterprise Social Media (ESM) use (Luqman et al., 2021; Y. Wang et al., 2021a, 2021b).
- Build organisational social capital and orchestrate corporate reputation online (Bentele & Seidenglanz, 2008; Ramos & Casado-Molina, 2021; Zarkada & Polydorou, 2014).
- Handle crisis communications and social media storms (Cheng, 2018; L. Wang et al., 2021a, 2021b; Xu, 2020).
- Market and brand the organisation and its products (Dwivedi et al., 2021; Rowley & Keegan, 2020) and create a seamless phygital customer experience through four distinct but interrelated strategies (Li et al., 2021)
  - social commerce,
  - social content,
  - social monitoring, and
  - social Customer Relationship Management (CRM).
- Identify and evaluate market and business data and intelligence.

Platform and software marketers and consulting firms, with the support of mass media hype, encourage managers and organisations to wholeheartedly jump on the digitalisation bandwagon. However, métron should guide the process of digital transformation, both with respect to its pace and level of commitment that is undertaken, and with respect to the way the people involved are treated.

Determining the pace of digital transformation and the degree to which an organisation is committed to it is a matter of accurately forecasting and measuring return on investment (ROI). Digital marketing, and especially marketing communications, for example, is being presented as affordable for even SMEs and micro-enterprises. This, however, used to be the case in the early days of social media platforms, but it is no longer true. The ease of growing audiences, the relatively large reach of a small advertising budget, and the overall cost-effectiveness of the digital marketers' services were nothing other than a classic market penetration strategy. As the numbers of users grew, and their levels of engagement with the platforms rose exponentially, the algorithms ensured that the digital advertising budget needed to be substantial to have an impact. The same applies to the cost and quality of market and consumer behaviour data, as well as that of general business intelligence which is available online. Valuable information posted by organisations who have not spent heftily on Google AdWords™, for example, is forever buried in the browsers' second pages, where no one ever looks.

Communicating in the Phygital Domain

Biased and fake news are fast becoming a threat to businesses. They are often confounded by deepfake scams and smear campaigns (Raval, 2024), which can even be targeted towards those of their managers who have a public presence. For example, Meta™ owner Mark Zuckerberg saw the most outrageous, but also some quite close to convincing, fake news about him, complete with deepfake photos and videos, plastered all over the internet when he announced that his platforms will no longer fact-check (Smalls, 2025). Some of these reputational threats can just be left to fizzle out, some require an investment of resources to be countered, others can prove detrimental. Correct assessment of the risks involved and proportional response are, to some degree, functions of the resources available to management. As a general rule, we see that the wider the exposure, the bigger the risks, and the costlier the response. Métron is the optimum approach to mitigating the risks and controlling costs; métron, both in the time and energy we invest in establishing an online presence, and in our expectations of the benefits.

Despite the need for caution, digitalisation is most likely the most powerful weapon in the hands of small organisations claiming their place in the market without the support of economies of scale, strong brands, negotiating power, and large marketing budgets. Just as the giant Goliath intimidated the people of Israel, so can large organisations intimidate contenders. No one dared to face the biblical Goliath until the shepherd David decided to confront him armed only with a slingshot and a handful of carefully chosen smooth stones. David met the giant Goliath and defeated him with just a correctly targeted stone that struck his forehead. As the hero's namesake, David Aaker (Aaker, 2020), suggested, competition by a business giant can be countered with four precisely targeted 'smooth stones': credibility, simplicity, personal contact, and a demonstrated higher purpose.

Digital communications, with their unprecedented reach, directness, and speed, can act as the sling that makes sure those that matter to the organisation, its stakeholders, perceive it as credible, appreciate its simplicity, enjoy the personal contact with its people, and embrace its higher purpose. It is critical, however, to remember that a long-lasting perception of credibility cannot be bought by flooding the World Wide Web with paid messages. It is built by repeated stakeholder interactions which are personalised, open, straightforward, transparent, honest, consistent, and empathetic. The 'your call is important to us' message people get when they call a corporate helpline is rendered insulting by a long wait or a dropped line. A sincere greeting and a smile can never be successfully substituted by a chatbot, no matter how fast and accurate.

In every transaction episode, various forms of value are exchanged (such as products, services, ideologies, experiences, money, time, attention) alongside the technical and economic information necessary for the transaction to be completed. Additionally, there is always a social dimension, which may range from formal greetings and farewells to more personal interactions, including the exchange of information about the personalities, lives, and values of the individuals involved. The more frequently and intensively the parties to a transaction communicate, the closer their relationship becomes, leading to mutual adaptation. In many cases, such interactions result in the institutionalisation of communication codes and, especially in interorganisational relationships, the standardisation of processes and customisation of products. Close and recurring relationships, eventually, develop their own characteristic atmosphere comprising the dynamics of power and dependency, level of collaboration, degree of proximity, and a set of mutual expectations (Ford, 1990; Sood & Pattinson, 2012).

Organisational sustainability and profitability rest on the mutually beneficial relationships the organisation has managed to build and sustain with its stakeholders. The credibility, simplicity, and personal contact required for this cannot be faked. Neither can virtue and a higher purpose. Contemporary customers, particularly Millennials—the largest and most ethically driven generation (Chatzopoulou & de Kiewiet, 2021)—seek relationships with organisations that pursue admirable and respectable purposes, such as philanthropy, fair trade, social justice, and environmental protection. Ethical consumption is of greater importance to them than price (López-Fernández, 2020). The consistent communication of an organisation's commitment to a meaningful mission, such as environmental protection, constitutes a competitive advantage, even for small organisations operating in markets dominated by large players, provided that the message and its timing are appropriate. At the same time, greenwashing is increasingly more often detected and penalised by consumers (Knight et al., 2022). So, in corporate communications, honesty is the métron.

For large organisations, the same 'smooth stones' applied to marketing communications can counter the negative effects of being perceived as too powerful, monolithic, impersonal, cold, and distant. This, however, does not solve the problems created by the fact that some organisations are actually too large for social structures to handle them, too powerful for political and legal structures to control them, and ultimately, instead of existing to serve the societies that host them, becoming detrimental to them. The effects of private enterprises exceeding métron have been summarised as follows (Putnam & Garrett, 2020):

> Corporate conglomerates are replacing local and craft economies in almost every sector, including agriculture. America's rugged individuals struggle against the loss of identity, autonomy, and mastery as they are subsumed into the anonymous labor of hyper-consolidated corporate machines and forced to pool meager wages to make ends meet. Corporate monopolies have hoarded profits and gained unrivaled economic influence through a wave of mergers. Because of corporations' outsized power, workers' leverage has eroded, and capitalists cite their responsibility to shareholders and market forces as justification for keeping pay low. Corporations search at home and abroad for ever-more-vulnerable populations to employ at ever-lower wages.

Managers are taught how to seek and exploit tax and labour law heavens, or how to negotiate and implement mergers, acquisitions (M&As), and various types of crossholdings to evade regulatory oversight by their sheer size and power. Bringing and keeping an organisation within the human métron, however, is not usually part of the business school curricula. It has been argued (Putnam, 2001; Putnam & Garrett, 2020) that the erosion of social capital, decreased levels of trust, rising inequalities, and widespread pessimism commonly observed in contemporary liberal capitalist societies are the result of the rifts with tradition. We seem to forget or brush aside knowledge acquired through human experience and thought over centuries, if not millennia. So, we here propose that, to achieve the mindset regeneration future managers need to bring about sustainable prosperity for all, we need to look into the treasure trove of our collective heritage for inspiration.

2.5 Building a Better Future by Utilising Traditional Approaches to Restoring Métron

The idea of keeping what was good and useful from earlier phases of human development and incorporating it with new ideas, technologies, and forms of social organisation to serve humanity is not new. Business practices and their tools, such as management, marketing, and finance, are products of the societies that they serve; they reflect their dominant ideologies and prevalent views of the world and its working and are intrinsically tied to the people who practice them and their thought systems. Just as the printing press and the First Industrial Revolution moved societies and organisations into modernity, electronics, telecommunications, and the Third Industrial Revolution, the Digital Revolution, brought them into the postmodern era. Theories such as relativity, quantum physics, topology, and fractal geometry shattered the belief in objective reality and initiated people into the concept of hyperreality, an agglomeration of socially or individually constructed subjectivities, stories circulating through media which carry no truth other than that of the time, place, context, and people who construct and reconstruct them. The self was replaced by a multiplicity of fluid and transient images, roles, and personas. In the postmodern context, communities and their organised value systems were shunned. The individual was king, ruling like a toddler demanding the immediate satisfaction of every whim, existing through conspicuous consumption. The affluence, relative peace, and stability of the post-Cold War era allowed societies to revert to an infantile state.

Enter multiple economic crises and their resulting mass impoverishment, and the postmodern existence is no longer financially possible for the many. Then COVID-19 comes to remind twenty-first-century people that the global village is far too small for unadulterated egoism. Finally, wars and the threat of wars in regions that were thought to be secure or where hostilities were thought to be contained amply demonstrated the importance of having and respecting organised value systems such as the distinction between combatants and civilians, the prohibition of war crimes and mass punishment by starvation, the universal application of the rule of law and the right to a fair trial for everyone, including leaders of terrorist organisations (at least before they are proclaimed heroes of revolutions against dictators who have, however, been repeatedly elected), human rights, and upholding referenda and election results, even when those in power do not like them. In the VUCA framework of the Fourth Industrial Revolution, people stand apprehensively in front of their lives turning into the science-fiction movies they grew up with. Emerging technologies, like artificial intelligence (AI), robotics, the Internet of Things (IoT), blockchain, and biotechnology, blend the physical, digital, and biological worlds into an exciting but largely confusing environment. It has been argued that we are fast entering the era of Transmodernism (Burns, 2015).

Transmodernism reclaims premodern values such as truth, beauty, goodness, trust, integrity, and harmony, alongside a renewed focus on wisdom, spirituality, and simplicity. It promotes a hopeful and balanced understanding of (a) the self through its familial and societal relationships—both spiritual and interpersonal—which play a significant role in shaping identity and (b) the complexities of the human experience through the ties that bind individuals and their values with other people and their shared values. It advocates the restoration of connections between the self and society by emphasising integration rather than rejection, community engagement rather than egotistical individualism, mindful integration rather than whimsical fragmentation, and substance over image.

No era exists in isolation of the others, and their boundaries are not clear-cut and impermeable. Just like in some parts of the world one can still find premodern societies, so the postmodern emphasis on picture and sound brings fundamental elements of premodernity into the digital world. So, transmodernism fuses elements of other frameworks (the solidity of premodernism, the progressiveness of modernism, and the scepticism of postmodernism) and integrates their key insights within a broader context acknowledging interconnectedness and redressing the alienation and despair of the permacrisis.

In this section, we briefly present some of the traditions that practically employ métron for the benefit of individuals and the societies that practice them. They can be incorporated into personal daily practices to improve quality of life and reduce the negative social and environmental footprint of individual and household consumption practices. Most importantly, they can be woven into the fabric of organisational higher purpose and promote environmental and social sustainability. Thus, one person, one household and one organisation at a time, societal practices change, and the course of humanity is set towards sustainable prosperity through the regeneration of mindsets.

2.5.1 Temperance—Learning from the Scandinavians

We have already introduced the Swedish concept of *Lagom* (Knowles-Dellner, 2017), literally describing the happy state of having neither too little nor too much. By focusing on what truly matters, lagom advocates finding contentment in mindful living within métron, specifically:

- maintaining balanced work and life commitments,
- seeking harmony between the different aspects of personal, organisational, and communal experiences,
- living within one's means, using resources wisely, and appreciating what you have rather than constantly seeking more (thus contributing to environmental sustainability),
- cultivating a community-oriented mindset, valuing collective well-being and fostering collaboration over competition (thus contributing to social sustainability).

Closely related to lagom is the Danish concept of *hygge* (Lubowiecki-Vikuk et al., 2021; Wiking, 2017)—a word sounding much like the Greek word for health *υγεία* (hygeia)—literally meaning the joy of cosiness, care, calmness, and collective welfare. While lagom is more of a value orientation and a process, hygge is a "desired end-state" resulting from an individual's "integration with society, as it embeds a personal feeling in relation to conviviality" and an alignment of one's goals within "the realm of social contexts (Mahajan et al., 2023)". Lagom and hygge are cornerstones of the Scandinavian culture which influence socioeconomic organisation in multiple ways. It has been argued that it underlies common practices such as generous salaries, long holidays and breaks during the working day, and extended parental leave, which facilitate work-life balance, increase productivity and innovation, and reduce unemployment at the country level. It is directly linked to health as people have "time to spend in the kitchen, preparing more food from scratch … [so that] eating well … is taken for granted (Knowles-Dellner, 2017)". So are exercising in the fresh air, playing, and learning in a relaxed environment which avoids rote learning and grade-hunting. These practices are underpinned by the values of an overall sociocultural framework which fosters creativity and encourages exploration. In turn, creativity and exploration increase innovation and productivity in organisations and, ultimately, in the economy.

Scandinavian design, cooking, architecture, and art express the beauty of moderation and simplicity and promote métron in all aspects of life. For example, the higher purpose underpinning Swedish brands which have become household names all over the world has influenced aesthetics and consumption patterns on a global scale. This demonstrates the ability of corporations to empower people in a way that makes a positive change in both their private lives and, ultimately, in the world. Moreover, the six 'Denmark values', namely 'Cosy', 'Unpretentious', 'Design', 'Talented', 'Oasis', and 'Free', which formed the basis of the country branding effort of the late 1990s, have guided not only state policy but also the business and marketing practices of a broad range.

There is a lot of academic literature (Lubowiecki-Vikuk et al., 2021) showing a switch of consumer behaviour away from materialism and towards lifestyles of métron, such as lagom, which "strongly underline concern for the environment, one's health and that of other people, and active leisure". There is also evidence of rising numbers, and increased spending power, of responsible consumers who support fair trade and local products and use social media to chastise products which are not compliant with quality, safety, and environmental standards. These responsible consumers expect the corporations they patronise to share their commitment and actively espouse and promote their values. They are the future managers, and, with the power of their positions, they can both alter the mindset and practices of their organisations and educate consumers to appreciate and actively espouse métron.

2.5.2 Connectedness—Learning from the Japanese

From the other end of Eurasia, the islands of Japan, come four key concepts which define the métron of how people relate to

- nature and artefacts
  - *mottainai*, the disdain for waste,
  - *kintsugi*, the celebration of resilience,
- people and groups
  - *sanpo-yoshi*, the three parties of a dyadic relationship
  - *kata,* the way and the human being.

We would like to make it clear that the discussion of these concepts is not an attempt to contribute to the large volume of literature which became world famous as *nihonjinron* (日本人論, a combination of 日本 (nihon)—Japan,人 (jin)—both person and people, and 論 (ron)—theory), or its specialised branch related to the Japanese economy (*nihonkeizairon*). Nihonjinron became popular after WWII and burgeoned after the success of Japanese exports in the USA and Europe in the 1970s, for obvious reasons. It comprises books and articles from the perspectives of anthropology, sociology, psychology, history, philosophy, linguistics, law, economics, and even the natural sciences, dealing with Japan's uniqueness and comparing its culture, socioeconomic organisation, and business practices with those of Europe and North America. From *The Chrysanthemum and the Sword* (Benedict, 1954) and the concept of *Mendaka* (目 (me)—eye and 高 (taka)—high, which describes the hierarchical nature of groups) (Nakane, 1973), to attempts to link Japan's historical and cultural roots with its modern identity (Tasker, 1988), there is a common thread running through nihonjinron, the idea of absolute 'Japanese uniqueness' expressed in five non-negotiable tenets:

1. Absolute singularity—there is no other race or nation like the Japanese in the world.
2. Linguistic singularity—the Japanese language contains words and phrases that are impossible to translate.

3. Racial uniqueness—the combination of genetic characteristics of the Japanese is different from all other peoples. This is linked to the conceptualisation of the Japanese people as a totally homogeneous population.
4. Homogeneity—known as *kokutai* (国体), which is translated as the country's sovereignty, polity, identity, and character—is the foundation of Japanese nationalism. It describes the totality of a unique cultural and political organisation of the people and paternalistic state around the spiritual and political leadership of the Emperor as the Father of the nation (Dale, 1990).
5. Ahistorical essentialism—the basic character of the Japanese originated in prehistoric periods and remains unchanged in the course of history irrespective of circumstances. This also ties with *Kōshitsu* (皇室), the Imperial House of Japan (Kōshitsu), the world's oldest continuous hereditary monarchy, which is believed to have an uninterrupted lineage all the way back to the Emperor Jimmu in 660 BCE, and certainly has had so since the sixth century CE Emperor Kinmei.

This conceptualisation is not without its challengers (Dale, 1990; Reading, 1993; Rear, 2017; VanWolferen, 1989), of course. However, fascinating as nihonjinron and its rebuttals might be, the point is that the key processes we have been experiencing in the late twentieth and early twenty-first centuries are that of cultural transnationalism (Iwabuchi, 2002) (i.e. the sharing, adaptation, and integration of cultural practices, values, and ideas over national boundaries and across different societies) and glocalisation (Robertson, 1992) (that is, the effects of the simultaneous universalising and particularising of cultural elements and global influences on local cultures and individual and collective identities). In this context, we believe that irrespective of the force of the Trump MAGA agenda, or the resilience of the World Economic Forum (WEF) global governance ideology (which was, ironically, meeting in Davos-Klosters just as Donald Trump was taking office (Markovitz, 2024)), cross-border exchanges and cultural interactions are not likely to cease. If anything, the cross-fertilisation of cultures within the framework of global interconnectedness, which no tariffs, border fences, and blocking of TikTok™ or RT News™ can reverse, will continue to foster hybrid forms of cultural expression and to facilitate the creation of transnational communities which have an affinity with and references to multiple places, cultures, and traditions. Future managers are such a community. Their individual and collective ability to bring about positive change in the world relies not only on their understanding of each other but also on their skilful and conscious enhancement of themselves and their practices through the infusion of this understanding. In this spirit, we discuss each one of the concepts and their function in the context of finding and restoring métron.

The Ethics and Practices of Zero Waste

In Episode 3 of the TBS/Netflix series *La Grande Maison Tokyo*, the search for game with which to participate in a cooking competition brings the owner and chef of a newly established *haute cuisine* restaurant to a hunter's lodge in the mountains. The hunter explains why his meat is the best in the world: "I believe it's respectful to eat the meat as deliciously as we can", he says, then turns away, bows and claps his hands, and bows again in front of the little shrine he has in his workshop, like people do in temples. He refuses to sell them his game and explains, "We receive life from nature. I can't give my meat to someone who wants it for selfish reasons". Later, he tastes the chef's cooking. As they discuss the food, the chef says, "As chefs we must make the most delicious meals from the life we take. Chefs exist for that purpose. We're just doing what we should as chefs". Finally, the hunter agrees to supply the restaurant with game, mushrooms, and herbs. "Please make the best dishes with the life I've taken" he admonishes them again.

Thanking the slain animal for its sacrifice, doing justice to its life by cooking it to the best of your ability, and eating it respectfully are demonstrations of the fundamental value of *mottainai* (勿体無い—translated as waste nothing, but uttered with scorn and indignation, implying that the practice is impious, even sacrilegious). It is not just a matter of "mere frugality or conservation; it is a religious concept" (Morita et al., 1988). Above all else, it is a matter of respect to nature and the resources it provides. The belief that whatever is available to people is a sacred trust to be used wisely, and that waste is irreverent is central to Japanese culture since time immemorial. Everybody is equally and indivisibly responsible for taking only what is absolutely necessary and treating it with the utmost respect and care. This attitude also guides Japanese industry. As the co-founder of Sony, Akio Morita, wrote (Morita et al., 1988), waste "carries the connotation of a sacrilege against heaven". The disdain for waste is evident in the Japanese approach to concepts such as luxury (贅沢—zeitaku) which connotes waste and reproachable extravagance or shameful indulgence (Hanley, 1999). This deeply ingrained attitude is, of course, as diametrically opposite as anything could be from the Western concept of conspicuous consumption (Veblen, 1998/[1899]), the practice of going to extremes to display wastefulness in the use of objects and time as a sign of inflated pecuniary status, which has prevailed among not only the high but also the aspirational classes since, at least, the turn of the twentieth century.

The concept of mottainai extends beyond the obvious to Westerners, such as, for example, energy saving, efficient utilisation of natural resources, and recycling, and reaches into every aspect of the private lives, industry, and economic structures of Japan. The oil crisis of the 1970s is a case in point. At the time, Japan was solely dependent on imported oil and its products were widely perceived as cheap copies of American and European ones. This reputation was systematically cultivated by the media and the entertainment industry in the Western world, especially in the USA. A typical example is Maxwell Smart's (Agent 86) phrase "Don't be fooled by a cheap Japanese imitation", which passed from *Get Smart,* the cult satirical spy television series of the 1960s, by Mel Brooks and Buck Henry, into the vernacular where it can still be found as a surreal joke. Instead of being catastrophic for the Japanese economy, the oil crisis catapulted Japanese exports and permanently changed the image of Japanese products from cheap to smart. It brought Japanese household appliances and cars which were swiftly redesigned specifically to save energy to the international spotlight and into European and American households almost overnight. The redesign was a result of a concerted R&D effort by universities, numerous industries, from steel manufacturing to electronics and shipping, and companies under the guidance and with the support of the government and its specialised agencies. During the embargo, the entire Japanese production base assumed the responsibility of finding ways for reducing energy consumption. This approach was not some departure from the norm. It was simply the self-evident reaction for both the establishment and the average Japanese person: assuming responsibility for finding a permanent solution to a temporary problem. The stance of the average Japanese person is key to such processes. Mottainai is a way of looking at the world and one's place in it. It is also a way of life focused on avoiding the sense of profound regret over wasted or inappropriately used resources, and the sadness of one's skills, time, energy, and opportunities failing to be developed to their fullest potential.

At the household level, this mentality has deep and timeless roots and has led to ever-improving and often innovative products that serve the real and psychological need for economy while incorporating the latest technology. The Japanese toilet, with the little faucet and basin on top of the tank, is the most characteristic example of absolutely no waste. You wash your hands under the faucet, and the used water flows into the tank and is used for flushing the toilet. The heated electronic toilet seat, about which every Westerner who has ever visited Japan has a funny story to tell, takes care of personal hygiene in the most efficient way without using toilet paper or loading the laundry basket with dirty towels.

In the cold winter months, instead of heating the whole house, the Japanese warm the human body and shut out the elements to maintain its temperature. Since ancient times and up to the, unimaginable for Europeans, integration of technology into the everyday life of people, the Japanese bathtub (お風呂 (ofuro)the honorific prefixお (o) and 風呂 (furo)—bathtub) is narrow and deep, so that one person, or the whole family, can sit comfortably (but not lie down) in warm water reaching up to the jaw. This design saves space, water, and energy by reducing heat loss from the water surface. People wash thoroughly before entering the bathtub, so the tub is filled once every day and the hot water (which usually has a temperature of around 40 °C to help relax the muscles and increase blood circulation so the body stays warm for hours afterwards) is used by the whole family and the house guests. Nowadays, water temperature and flow are controlled by complex electronic devices which take advantage of low power demand and efficiently accommodate the different schedules of the household members.

After bathing, the human body simply needs a small device, the *hibachi* (火鉢—literally the fire bowl), that warms the legs under the low dining table and a warm silk duvet to spend even the coldest night in comfort. The hibachi used to be a closed charcoal warmer but today is either an electronic warming table or an electric tablecloth-cum-blanket with timers and electronic thermostats. The idea of the *hibachi* was introduced from China around the eighth century but, from the time of its diffusion by the samurai and the aristocracy to the broader social strata in the Edo period (1603–1868) it kept evolving with time and technology. It has served as a tea kettle warmer, lighter for cigarettes, the portable cooking device of

Japanese soldiers during World War II, and it still is, in its original form, integral to the traditional tea ceremony. The hibachi was developed by the home appliance industry into the sophisticated table-top electric cookware which is now very popular around the world under the name *teppanyaki* (a western mixture of the utensil, the hot iron plate鉄板 (teppan), and the cooking method 焼き (yakki)—grilled).

The culinary tradition of teppanyaki also expresses the mottainai principle both in the sense of doing justice and showing respect to the sacrificed flora and fauna and in the context of saving water, energy, time, and effort. Thin slices of meat, fish, and vegetables are very swiftly seared on the very hot plate, all at the same time and with no fat or liquids other than that of the raw food. This saves time and energy and preserves all the nutrients and natural flavours. Moreover, there is no cooking smell emitted, so there is no need for exhaust fans which consume electricity or open windows which bring the outside cold into the house and might irritate the neighbours. Water and detergents, as well as time and effort, are also saved since everything is cooked in one utensil which is simply wiped with a piece of paper after use and is ready to be reused.

Finally, nowhere is there a starker contrast to the approach to full utilisation of resources between Japan and the west as there is in apparel traditions.

> Clothing took a minimum of material, certainly compared to Western dress, used an entire length of cloth with no waste, and could readily be adapted for anyone without resewing a garment. ...
>
> Kimono are made from one long, rectangular length of cloth cut into eight pieces; the pattern is the same for everyone, male or female, child or adult. Every inch of cloth is used with no waste. No matter how efficiently an article of Western clothing is cut, in fitted clothes there will be pieces from which nothing can be made or at least that are waste from a particular garment. One exception might be women's skirts, but they require so much more material than a kimono that they are scarcely as economical. There are no buttons or hooks or any other gadgets needed to hold the kimono on. The obi (the band or sash tied around the waist) is sufficient. This means that when the garment is put on, it can be adjusted for how fat or thin the wearer is as well as for how tall. In contrast, Western clothes are made to fit one person. If that person grows or gains weight, the garment no longer fits, and likely as not, no one else will be the perfect fit for it. Often even hand-me-downs have to be resewn, and in a used garment such reworking will likely show. ...
>
> Like clothing, footgear was standardized in Japan. The basic footwear was a sandal with thongs between the big toe and the second toe to hold it on. The poor and travelers wore geta (wooden clogs) or waraji (straw sandals), which were not only readily and rapidly produced from materials around the farm, but almost anyone could wear anyone else's. Thus, they could be produced without the regard for size and fit necessary for Western boots. Straw footgear wouldn't last long, but it was cheap, and one could readily buy or make more. Such footwear didn't provide the protection that boots did, but then, one didn't get bunions or corns from a bad fit (Hanley, 1999)".

Embracing mottainai guides the Japanese approach to new product development (NPD). It is common practice to create detailed catalogues of product designs listing all their components. This way, they simply use parts of previous products to create a new product. Existing components and parts can comprise up to 90%

of the final product, thus significantly reducing manufacturing costs and design and production time (Fraser & Zarkada-Fraser, 2001). Changes to the component designs can only be made when there are concrete indications that they will lead to an even greater reduction in resource use. To safeguard mottainai, Japanese industrial designers work in teams, with only one part of the project assigned to each designer, instead of one person undertaking a whole project, as is common practice in the USA (Fraser & Zarkada-Fraser, 2001).

The reuse of designs and parts in modern Japan has been extensively facilitated by the industrial superstructure. Historically, the Japanese economy has been structured around large and close-knit corporate groups. Up until WWII, the dominant force was the *zaibatsu* (財閥—literally the wealth clique), the monolithic, family-controlled conglomerates like Mitsubishi and Sumitomo, which still exist today as brand names of corporate groupings. Even though the Allied Occupation forced the zaibatsu to dismantle, they soon regrouped into the keiretsu (系列—system of connections) which, although weakened after the burst of the bubble economy in the 1990s and recently becoming increasingly corporatised, still remain as networks of interlinked companies, both in terms of cross holdings and, in the spirit of mottainai and efficiency in resource conservation, in terms of bank, supplier, manufacturer, and merchant relationships, such as the Toyota fully vertical supplier network. No matter what form they take, it is an indisputable fact that enterprises linked by such close relationships participate in joint research efforts and distribute the tasks and costs of developing materials, processes, and products in the most efficient ways possible; they coordinate their procurement and subcontracting, exchange ideas and know-how, and enjoy high levels of mutual adaptation, trust relationships, and friction-reducing institutionalisation, thus increasing their bargaining power (Aoki, 1988; Kodama, 1995).

The logic of the repetitive part of a design finds its fullest and most impressive application in the construction industry (which we discuss in detail in *A case in point: the Japanese construction industry*). Japan relies heavily, and more than any other country, on industrialised prefabrication and strict standardisation which is based on interdisciplinary research and development (R&D) carried out mainly by the research institutes of the large construction firms. The key objective is not to invent unique approaches but to achieve the most effective fusion of ideas, materials, technologies, processes, and aesthetic and cultural approaches from around the world (Fraser & Zarkada-Fraser, 2000). Even in cases of unique, non-repeatable designs, the concept of mottainai dominates along with other elements of Japanese culture such as teamwork, innovation, and full utilisation.

The story of the Kobe Paper Church is a typical example. The church was built in only five weeks in 1995 by 160 volunteers, using materials donated by industry, to replace the Takatori Church, which was destroyed by fire following the Kobe Earthquake. Shigeru Ban's plan borrowed the elliptical shape of the Benigni churches. The structure rested on 58 paper tubes, commonly used for moulding concrete, and was surrounded by wavy polycarbonate foil that is usually used as packaging material and is transparent, thus eliminating the inside/out separation.

(You can see stunning photos on the website of Shigeru Ban Architects. https://shigerubanarchitects.com/works/cultural/paper-church-kobe/). In 2005 the church was dismantled, and the materials were sent to Taiwan, where it was reconstructed. Of course, this original application of materials is not an isolated example but the norm. In Japan, the multifaceted dimension of materials and the diversity of synthetic elements that may have been designed and manufactured by seemingly unrelated industries and perform many different and unrelated roles are taken for granted.

Finally, mottainai has been successfully integrated with state-of-the-art techniques like scorecard-based evaluation and monitoring and the life cycle assessment (LCA) framework to identify, measure, and target greenhouse gas (GHG) emissions in collaborative automotive and pharmaceutical manufacturing networks to achieve sustainable supply chain operations (Adhya, 2024). Thus, it can be a cornerstone of future management and a catalyst for prosperity for all, now and in the future.

Knowledge Is a Public Good

Closely related to the concept of mottainai is the understanding of knowledge as a common, almost public good. It is something to be shared and utilised as widely as possible so that not even a thread of it is wasted. The concept of technology as a closely guarded secret which must be protected at all costs, especially in the courts, is alien to the typical Japanese. For Sony, for example, maintaining first-mover advantage is the active way of dealing with what they see as their inevitable responsibility.

> We have poured billions into our technical laboratories for R&D, and others have watched us and taken advantage of our up-front investment by moving into our field after we have pioneered it, but we cannot change this. I have no intention of complaining to anybody. We have boasted about our capability in these crucial areas, and now the responsibility is on management in the next fifteen years to see what we can do. We would have to be asleep if we did not produce any fruit from such a lush orchard. (Morita et al., 1988)

Globalisation has led Japanese businesses to litigate against overseas competitors for patent infringement and to protect intellectual property rights (IPR). These cases, however, are few and notable rather than the norm. In the domestic market, and within their supply chain networks, the focus on mottainai, maintaining harmonious business relationships, and resolving disputes through negotiation and arbitration in the domestic milieu has not shifted. Designs and know-how are shared as part of the conceptualisation of the economy as a port: when the tide rises, all ships go up (Holden & Burgess, 1994), which, in a sense, is the antipode of the inevitability of "the perennial gale of creative destruction (Schumpeter, 2013/[1942])".

Unlike the western fixation with impressive radical innovations, which are associated with having a strong leader, such as a visionary or a very powerful company in an industry, Japan's success has been based on technological fusion, which requires collaboration within and across industries, even between directly competing firms. There are, of course, variations in the degree of sharing of knowledge and the roles assigned to each of the members of the supply chain network, but the key idea is that there should be as little siloed processes as possible and as much collaboration as financially and technically feasible so that labour, responsibilities, and gains are divided proportionately to the capacity and needs of participants (Kamath & Liker, 1994). Unlike what happened in the USA with the semiconductor revolution, which destroyed many businesses but gave rise to a few that grew rapidly, in Japan the explosion of mechatronics demonstrated that high technology diffusion rates, such as the one prevalent in the industrial robots sector, were linked to overall greater growth throughout the economy (Kodama, 1995).

Finally, studies of Japanese companies abroad (Holden & Burgess, 1994) have shown that knowledge sharing takes place not only between Japanese firms in a supply chain network but also with foreign firms participating in it. These relationships, which begin very formally and with a limited agenda, not only evolve into openly and actively collaborative ones but, eventually, progress to joint product development and technology transfer. Most interestingly, they come to the point where the foreign firms enjoy the mutual adaptation and the institutionalisation of information, intelligence, and knowledge flows, they feel trusted and important and, eventually, embrace the customer-centricity and participatory quality management systems of their Japanese counterparts, even when the Japanese supply chain participant is relatively small in financial terms.

Built-in Obsolescence Need Not Be by Tomorrow Afternoon

Back in the 1930s, the pioneers of marketing (Hess, 1935) described the "new world in the making" they were heralding as "a consuming society" in which "old habits begin to break and tradition and precedent give way to the enthusiasm of new experiences". They saw their role in it as helping to "build business around the structure of the individual… [and] make nature and human nature co-operate through the business structure". They conceptualised the "consuming society" as follows:

> It accepts the challenge of speed in all relations as a measuring rod indicative of our spiritual capacity and character to transcend time and spatial limitations, thus consciously creating time and energy to do other things. By doing this, alone, can we foster the maximum consumption of goods.

They dreamt of their becoming "engineers of human efficiency in relation to the wholesome consumption of goods" through which

> not only is the individual given a sense of dignity and awareness of order in his (sic) life, but the environment is constantly being enriched in terms of utility and art. To put the thought in another way, if the world is to be a more beautiful and livable world one hundred years from now, in a mass production world, it cannot be a 'hodge-podge' world as now confronts us. It will undoubtedly be a world dominated by functional living wherein human energy, speed and creative living will prevail. (Hess, 1935)

A challenge stood in the path of achieving this dream: "to discover how to get Mr. and Mrs. Consumer to consume (Hess, 1935)". Just three decades later, our intellectual forefathers had managed to, not only successfully address the challenge, but also to turn 'Mr. and Mrs. Consumer', and their offspring into *The Waste Makers* (Packard, 1960). The 'hodge-podge' world of the 1930s had turned into the 'throwaway society' (Packard, 1960), which is the ever-expanding and dangerously all-encompassing norm ever since. We should not wonder how the dream turned into the permacrisis, if not a nightmare.

It is simple: the tents, climbing equipment, food wrappers, plastic bottles, and miscellaneous other waste which turned Mount Everest into the "world's highest garbage dump (National Geographic, 2025)" did not fall from the few remaining trees. It is explainable: we consider no part of the universe as off-limits to various forms of exploitation, such as mass tourism or resource extraction, and we value convenience over durability, thus causing an explosion in the consumption of disposable, single-use, and non-repairable products. What is worse, we do it with glee and take pride in it. For example, we celebrate consumer culture (Kravets et al., 2017) and pore over consumer stories seeking meanings in brands and "the symbolic boundaries that structure personal and communal consumer identities" in product symbolisms and consumption rituals (Arnould & Thompson, 2005). It is characteristic of our attitude that in a 551-page-long *Handbook of Consumer Culture* (Kravets et al., 2017) the word 'waste' and its variants appear in the passing just eight times (most notably just twice in a chapter on sustainable consumption), 'repair' only five (in the context of quirky hobbies, the peculiarities of Soviet life (Gurova, 2017), types of service jobs, and consumer protection legislation), and 'reuse' is totally absent.

Hope that seemed to be stemming from common sense initiatives such as the sharing economy and recycling has proved to be false. To evaluate the efficacy of the "emerging socioeconomic groundswell; the old stigmatized C's associated with coming together and 'sharing' – cooperatives, collectives, and communes", which were "being refreshed and reinvented into … Collaborative Consumption", one should ask the people of Athens and Barcelona, who cannot find suitable accommodation in their own home towns, if they think that in the "Airbnb community … rooms can't be a commodity (Botsman & Rogers, 2011)". As for recycling, according to recent studies (Ashna & Bandyopadhyay, 2024; Lahdili, 2015; Mehra & Sarkar, 2024; Nuryana et al., 2024), it costs too much and has too little effectiveness as a solution to waste management, especially with respect to industrial materials such as plastics.

The cornerstone of the throwaway society is planned obsolescence, the design and manufacturing of products with short lifespans and limited-to-no repair potential to encourage frequent replacement. The practice was documented around the 1960s (Galbraith, 1998/[1958]; Packard, 1960) and has since been repeatedly demonstrated to be spreading fast, throughout all sectors of the economy (including necessary products such as mobile phones, household appliances, and clothes), globally (Wieser, 2016). The key issue with planned obsolescence is that it is "a deliberate decision by suppliers (Wieser, 2016)" to use all sorts of means—from design to choice of materials, components, and software, and from advertising to withholding after-sales service and support—to ensure that their product will be replaced at a pre-determined point in time, which, in the twenty-first century, seems to be coming closer and closer to the time of purchase, thus fuelling a vicious cycle of overconsumption and waste and resulting in the societal normalisation of wastefulness. Future managers cannot continue to contribute to this practice. There will be no room for sustainable prosperity for anyone, and the galaxy will soon be too small to contain the waste we produce, if we do not immediately re-examine the morality as well as practicality—at least in the long run—of planned obsolescence.

We propose that we can regenerate our mindset by studying the Japanese art of kintsugi (金継ぎ, literally gold joinery). In its fundamental form, the kintsugi artist repairs cracked and broken crockery in a way that turns would have been its demise to the next chapter of its timeless narrative. Technically, pieces are joined together, missing parts are filled in, and cracks are covered using molten precious metals, lacquer mixed with gold, or metal staples to create a new object of increased worth and enhanced beauty.

Philosophically, kintsugi embraces the results of mishaps which are skilfully and carefully addressed as an object's unique and valuable features. It celebrates connection, continuity, the inseparability of the parts from a cohesive whole, and the value of timeless reintegration while constantly reminding the user of the inevitability of change and the inherent impermanence of all things human. Finally, instead of hiding flaws or imperfections, kintsugi integrates them into a new form in a way that highlights the beauty of transformation and the importance of resilience. These ideas are conveyed by the second character (継). It is made up of the thread radical (糸), which signifies the act of joining, like fabric with threads, combined with continuation (⺇) and the character for tree (木), which symbolises growth and life and, by extension, transformation and sustainability. The latter is one of the most important characters in the Japanese written language. It has a trunk, the vertical line, branches, the horizontal line, and at the base two spreading lines for the roots. Its incorporation within a character about connection reflects the value Japanese people place on their connection to nature and the Shinto tradition, which sanctifies trees for housing spirits. On a personal level, kintsugi is a method for dealing with life's adversities. It reminds us of the need to continuously work on healing and recovering. It also highlights the fact that our wrinkles and scars are to be cherished as the physical markers of the life choices that made us what we are.

Dyadic Relationships Must Always Have Three Satisfied Parties

The business philosophy and practice of *sanpo-yoshi* (三方良し—usually translated as three (三) way or persons (方) satisfaction (良)) originated from the thirteenth-century itinerant merchants of the Omi region, known as the *Omi shonin*. By the Edo period (1603–1868) the Omi shonin had established trade routes and markets all over Japan, and they were the main purveyors of specialist goods such as lacquer bowls, silk, and medicines. In the middle Edo period, the Omi shonin

Nakamura Jihei (中村治兵衛) elaborated sanpo-yoshi in his will with these words to his descendants: "Always put the customer first. When visiting other provinces, do everything you can to ensure that the people there feel good about using the products you bring (The Association to Spread the Concept of "Sanpoyoshi" Worldwide, 2010)". Thus, he firmly and formally introduced the necessary third party in the dyadic relationship between the merchant and the customer: the people of the provinces.

Omi shonin fully understood that they could thrive only through the approval and support not only of their customers but also of the communities they served. To ensure societal approval, they were involved in a wide range of community activities from funding large-scale infrastructure work, such as roadway and bridge construction and repair or the paving of dirt roads, to physically participating in routine facility maintenance, such as disposal of market refuse or regular road washing (Hanley, 1999), and subsidising festivals and social services, such as facilities for those in need or caring for orphans. Moreover, having established efficient transportation and logistics systems for their goods, Omi shonin naturally became the bearers of messages and news between people, communities, regions, and the establishment.

Their survival and prosperity depended on their access to timely and accurate information about trade routes and opportunities, products and their makers and prices, consumer demand and tastes, market and social conditions, and trade regulations and taxes. To ensure the steady flow of reliable intelligence they needed, they cultivated relationships of trust and relayed messages within a vast network of people over great distances. Thus, they became commercial information brokers, research, advertising and public relations agents, private, communal, and institutional message bearers and a public news agency. For centuries, they offered a valuable, honest, and reliable community service for which they were paid with the reputation, engagement, reach, and trust they needed to get more information and to keep their businesses growing, their customers satisfied, and themselves connected with their customers and the communities they served. Over centuries of operating in a specific way and with a set mindset, customer satisfaction and community engagement became their way of life and became ingrained into the national business fabric.

Times have changed, but sanpo-yoshi is still the reigning mentality of Japanese commerce and the fuel of the longevity and international success of Japanese firms (Morishita & Tainaka, 2019; Nguyen et al., 2024; Shinohara & Kim, 2022). Adherence to the spirit of sanpo-yoshi means simultaneously pursuing benefits to the seller, the buyer, and the local community. It acknowledges the inseparability of the three and seeks harmony in their collaborative coexistence. It draws on the Confucianist emphasis on interpersonal and societal harmony and the Buddhist and Taoist traditions which stress balance and interconnectedness. Thus, sanpo-yoshi views commerce not as the pursuit of profit for oneself, but as an activity through which all persons and groups which are involved, or even remotely affected, will receive a balanced and fair share of benefits. Much more than a commercial practice and a business philosophy, sanpo-yoshi is a way of thinking about one's activities in relation to the others and the world.

The way we see the world is expressed through our cultural and personal choice of symbols to represent our ideas. So, we would like to briefly discuss symbolic representations of concepts, without any intention of turning this into a discussion on linguistics. Sanpo-yoshi is written in its simplified form (三方よし) not only on the website of the *Association to Spread the Concept of Sanpo-yoshi Worldwid*e (The Association to Spread the Concept of "Sanpoyoshi" Worldwide, 2010) but also in scholarly works aimed at an international academic audience (Nguyen et al., 2024; Shinohara & Kim, 2022). This form, which is more commonly used in the Japanese business vernacular, emphasises the inclusivity of the business perspective it advocates. The first character (三) is obviously just the number three. The second character (方) signifies both a person and people. It also denotes the way as in the method and the form, but also the direction (in movement) and the side (as a stance) one takes. This we discuss in detail in the section entitled *This is how we can do it: design, process, coordination, adherence, and support*. What is important, here, is that sanpo (三方) means both three parties and three-way.

We diverge from the norm and choose to use the traditional, more formal, dictionary writing of sanpo-yoshi (三方良し) not only because we are, after all, talking about traditions. Further to highlighting the historical continuity of the concept, this character configuration stresses the moral dimension of sanpo-yoshi and represents its depth and nuance most eloquently. Yoshi (良し) is usually translated for westerners as 'satisfaction' (The Association to Spread the Concept of "Sanpoyoshi" Worldwide, 2010), but in the Japanese language, the kanji (良) is used to denote what is good, right, and proper. It is also used to express success, excellence, positivity, appropriateness, and virtue. It has an emphatically moral connotation but also signifies the societal approval the virtuous enjoy. Thus, satisfaction is one of the outcomes of achieving practical, moral, and societal goals, but it is not all there is to sanpo-yoshi.

In the context of business, sanpo-yoshi is the integration of pragmatic, moral, and relational imperatives with their outcomes for sellers, buyers, and society within the transactional, immediate, and long-term horizon. These can be schematically represented in a 3X3X3 matrix (Table 2.1) for easy reference. However, this is just schematic as, in the practice of sanpo-yoshi, they are a coherent whole; a unity of thought, practice, and results.

At the pragmatic level, sanpo-yoshi focuses on the effort required to meet business objectives. Sanpo determines the scope of business decision-making to include society on an equal footing with the seller and the buyer. Thus, in modern terms, it forces consideration for all stakeholders. Yoshi is the guiding principle for decision-making. The good, right, and proper decision is the one that benefits all stakeholders equitably, causes no harm to any stakeholder, and maintains harmony among them and in their relationships with the organisation. It follows almost deterministically that the decisions the organisation will make under the sanpo-yoshi mindset will result in:

Table 2.1 A schematic representation of the imperatives of sanpo-yoshi and its outcomes for the three parties involved in all dyadic business relationships

Dimension	Imperative	Horizon	Outcome		
			Seller	Buyer	Society
Pragmatic	Success	Transactional	Profit	Satisfaction	Economic growth
		Immediate	Viability	Value	Progress
		Long-term	CLV growth	Security	Quality of life
Moral	Virtue	Transactional	Trust	Fair treatment	Peace
		Immediate	Loyalty	Safety	Sustainability
		Long-term	Growth	Customisation	Enrichment
Relational	Harmony	Timeless	Noone is disadvantaged		Sustainable prosperity for all

- functional, aesthetically pleasing, safe, and durable products,
- socially and environmentally sound production methods,
- fair pricing,
- empathetic, honest, and transparent marketing communications, and
- efficient and sustainable supply chain management practices.

The sellers work hard and succeed in bringing buyers the products they need at the price they can afford. So, they make a fair profit with each transaction. Buyers are satisfied, and they buy again; the sellers see their businesses become viable, and the buyers get value from their relationship with the sellers. In the long run, buyers feel secure that they will continue to have an uninterrupted supply of products that satisfy them. As transactions increase in number and value, the economy grows; more businesses come into the market and invest in creating better products and more efficient processes, so technological process is accelerated, and quality of life improves for all at the societal level.

It is rather obvious that what we are describing here as the pragmatic dimension of sanpo-yoshi is very much how capitalism was supposed to work: fairly and honestly, and within the tradition of virtuous capitalism, profit generation is dependent on production; business expenses in the form of wages and taxes are "being plowed back into the local economy, which then produces greater business demand, and so on (Callanan, 2015)". This, of course, is nothing like the brutality of rentier capitalism (Dannreuther, 2024; Fana & Villani, 2024; Sadowski, 2020; Standing, 2021) which the peoples of the west are experiencing in the twenty-first century.

Because in rentier and its sister forms of crony (Acemoglu & Robinson, 2012; Whig, 2012), platform (Sadowski, 2020), etc., capitalism, decisions are not inclusive, as in sanpo (三方), the inordinate economic gains of the few are derived from resource extraction, asset ownership, and rent collection at the expense of the many, the precariat (Fuchs, 2020; Standing, 2021), that is, the masses who have no access to factors of production other than their skill, time, and effort, no income security, and no input in the way capital or assets are managed (Fana & Villani, 2024). As there is no incentive for the rentier classes to invest into productive activities and innovation, the growth of the real economy is stunted and quality of life decreases. Moreover, excessive control is accumulated around platforms which control value chains globally and establish intellectual and information monopolies (Dannreuther, 2024; Sadowski, 2020). With increased power concentration comes the erosion of checks and balances and the weakening of institutions, such as democracy and rule of law (Standing, 2021). Thus, there is no mechanism for the equitable division of burdens and gains. The results are the components of the permacrisis: speculation, stagnation, volatility, environmental degradation, inequalities, debt crises, political lethargy, and social disconnect. Rentier capitalism is capitalism with no métron. We here propose sanpo-yoshi as a pragmatic métron of business-level decision-making which can restore the lost métron, one business at a time, even within the capitalist framework.

We propose that a key measurable incentive for the organisation-wide adoption of sanpo-yoshi within the current form of the capitalist system is that, on a pragmatic level, it can yield customer life cycle value (CLV) growth and better management of customer relationships in both contractual and non-contractual contexts, a task that has proved notoriously difficult (Singh & Jain, 2014) under the current egocentric and profit-orientated approach prevalent in most businesses. CLV is a critical metric of the total value a customer brings to a business throughout their entire relationship. Although based on historical data, it is forward looking (Fader et al., 2005) and factoring in not only revenue, profits, and acquisition and retention costs (Kumar et al., 2008), but also purchase quantities, interpurchase time, number of repeated purchases (Fader et al., 2005; Singh & Jain, 2014) and returns (Singh & Jain, 2014), upsells and cross-sells, referrals, and, now with the social media technologies available, even reputational gains. It is obvious that CLV growth is directly linked to business viability. Moreover, by accurately estimating CLV—something which is greatly facilitated by digital technologies, advanced algorithms, and the exponentially increasing availability of fine-grained customer and transaction data (Fader et al., 2005)—businesses can efficiently allocate resources for marketing, sales, and customer support, fine-tune segmentation, targeting, and positioning (STP), design and deliver engaging customer experiences, and thus increase CLV.

As with all metrics, however, CLV overemphasises the seller perspective and, by turning decision-makers' attention to customer profitability optimisation, encourages customer value extraction activities which largely ignore customers' value perceptions (Lemon & Verhoef, 2016). This is a slippery slope. For almost three decades (Kumar, 1999) we know that "relationship-oriented … firms achieve higher returns on their investments than do transaction-oriented firms … and add more value to their services over time". At the pragmatic level, good products and customer support can generate customer satisfaction, and doing so consistently makes buyers perceive value they receive from the sellers as adequate and feel secure in their relationship with the seller. For the same three decades (Johnson &

Auli, 1998), however, we know that customer satisfaction does not guarantee loyalty. The transformation of long-term satisfaction into loyalty does not take place in a social and emotional vacuum. It develops within a trust environment (Johnson & Auli, 1998).

Trust is earned through consistent and convincing demonstrations of virtue on the part of the sellers. The fundamental moral imperative in sanpo-yoshi—expressed by yoshi (良し)—is virtue, which in business transactions is expressed as.

- acceptance of serving customers and communities as a personal obligation,
- expression of gratitude for the custom, trust, and loyalty sellers enjoy,
- maintaining reciprocity, and
- unwavering commitment to fairness, truth, honesty, transparency, and empathy.

The customers of virtuous sellers feel fairly treated and safe in their relationships with them, and so they are willing to impart information which will get them the customised services and products they want. On the sellers' side, their customers' and communities' trust, loyalty, and mutual adaptation form the bases of their long-term growth. So, it is in the moral dimension that relationships are solidified enough to become a factor contributing to social peace, environmental and social sustainability, and community enrichment; in a nutshell, firmly adhering to the moral imperative of sanpo-yoshi makes the world a better place. Virtue is not some kind of self-sacrifice; it is not even an altruistic stance. It is a self-preservation choice.

Contemporary consumers are exerting increased pressure on corporations to behave themselves (Bray et al., 2011; López-Fernández, 2020; Pauluzzo & Mason, 2022). They also seem to join with their shrinking wallets and disillusioned voices the ranks of scholars advocating that actors need to refocus their activities towards raising mean market morality (Coffin & Egan-Wyer, 2022). As the permacrisis expands and deepens, we expect that tolerance of unethical and amoral business practices will diminish. Even without external forces, many companies, even some which were believed to be too big to fail (Nurisso & Prescott, 2020), have managed to "destroy themselves from within (Callanan, 2015)" through the relentless pursuit of shareholder wealth maximisation at the expense of all other alternative objectives, which compromises efficacy, effectiveness, and efficiency, and precludes the pursuit of excellence (Callanan, 2015). Most importantly, abject egotism leaves no room for harmony at the relational level.

Sanpo-yoshi is all about timeless harmony. When practiced wholeheartedly, it simultaneously ensures that neither party of the dyadic buyer–seller relationship is in any way disadvantaged and sets the basis of sustainable prosperity for all at the societal level. At the relational, but also the moral level, sanpo-yoshi is not, as some have argued (Nguyen et al., 2024; Shinohara & Kim, 2022), the same as or similar to corporate social responsibility (CSR) and environmental social governance (ESG) compliance. We believe that sanpo-yoshi is closer to Archie Carroll's (Carroll, 1998) original corporate citizenship conceptualisation as the voluntarily assumed responsibility of business to behave just as any other good citizen, that is, to:

- "Be profitable (carry their own weight or fulfill their economic responsibilities);
- obey the law (fulfill their legal responsibilities);
- engage in ethical behavior (be responsive to their ethical responsibilities);
- give back through philanthropy (engage in corporate contributions) (Carroll, 1998)".

To these we would add to:

- use their influence power (that is, resources, knowledge of communication techniques, and media access) to "influence people's views on responsible and ethical business (Shinohara & Kim, 2022)".

With the latter, we do not imply that the practice of sanpo-yoshi is to be publicly demonstrated as part of the marketing communications package, the way CSR is often paraded. Good deeds, care and compassion are, in the context of sanpo-yoshi, moral obligations of the individual and, as such, to be performed with discretion and honest empathy, much like the Biblical "*But when thou doest alms, let not thy left hand know what thy right hand doeth* (Matthew 6:3) (*Holy*Bible, 2011)".

Sanpo-yoshi has also been related to the game theory-based 'win–win' concept (Morishita & Tainaka, 2019). We here argue that in the framework of deep and personally experienced harmony in all relationships, the idea that the parties to the relationship allow room for loss is simply absurd. If anything, sanpo-yoshi is the "triple-win" (Yagi & Yamash'ta, 2022). The permacrisis has brought home the "need to investigate ways to avoid the greedy, unethical, and strongly selfish behaviors under capitalism without eroding the incentives for each agent to make efforts for future development (Yagi & Yamash'ta, 2022)". Stakeholder theory has been around for about 40 years (Freeman, 2010/[1984]), and although not a philosophically uniform system (Friedman & Miles, 2006), it has exerted significant influence on management thinking and practice. It is mostly discussed in the context of improving the environmental and social sustainability of business (Hahn, 2022) and as a strategic tool (Atkinson et al., 1997; Jones et al., 2018), especially for communication (Perrigot et al., 2012) and reputation management (Lee, 2012; Puncheva-Michelotti et al., 2014) on the basis of economic rationalisations (Blair, 1998). We are not alone, however, in stressing its inherent moral dimension (Beekun & Badawi, 2005; Carroll, 1991, 1993; Carroll & Buchholtz, 1996; Logsdon & Yuthas, 1997; McCarthy et al., 2012) and strong value and values orientation. To this we will return in the next chapter exploring value.

This Is How We Can Do It: Design, Process, Coordination, Adherence, and Support

The Aristotelian Judaeo-Christian tradition is born out of the harsh sunlight of the Mediterranean reflected over the white marbles of the Parthenon, the calm sea-water, the bright foliage and fruit of citrus trees, and the subtly moving sparkling sands of deserts and rocks. At midday, the shadows are totally black and very clearly defined shapes over a blindingly bright background. Our collective experience over the millennia is that of antitheses. Thus, the western world is black and white, right and wrong, heaven and hell, us and them, inside and out. We build solid structures with materials that are as hard as they can be, and we make them to stand out and attract attention. In the earthquake-prone areas, we build with stone, marble, concrete, and steel. We construct buildings capable of withstanding the forces of nature to keep the juxtapositional order which we have created as much as possible. We are Catholic or Protestant, belong to either the Ukrainian Orthodox Church or to the Orthodox Church of Ukraine (Skorkin, 2024), and we readily kill each other over it.

Yoshimi Fumiko, my Japanese friend, took me to Nikko, where we spent time with the Buddhist monks of Rinnô-ji and participated in the Autumn Festival rituals at the Tôshôgû Shinto shrine. Every Easter she would come with me to the Holy Resurrection Cathedral of the Japanese Orthodox Church in the Chiyoda district of Tokyo to celebrate Easter in earnest reverence and quoted Billy Graham, the American Southern Baptist minister, in our conversations. When she moved to Australia, she attended Mass in her neighbourhood Catholic church every Sunday but never missed Easter in the Brisbane Greek Orthodox Church of St George in Brisbane with me. It is no wonder, then, that in a country of about 125 million people there are over 200 million professed religious adherents while religiosity levels are falling, and agnosticism and atheism are at about 13%, which is roughly the same as in the UK and USA (Hardacre, 2025). The traditional Japanese house is made of locally sourced, totally biodegradable clay, bamboo, wood, rice straw, cotton, and paper. In a country where the ground shakes practically every day, the objective of the structure is not to withstand the forces of nature but to move with them and, if it fails to return to its upright position after the earthquake, to cause as little damage as possible to the people in and around it. Moreover, the whole purpose of construction is to be in unity with nature, to blend in and echo its materials, colours, and shapes. Internal dividers, the shoji, are sliding paper screens made from mulberry-tree bark. They allow spaces to be open or closed at varying degrees, they let the light flow through, and they allow those sitting on the one side to see the moving shapes of those on the other side. They, and the rest of the world inside and out of the house, are a harmonious synthesis of vaguely outlined, floating shapes. They make up a world comprising muted sounds and shades of grey, one of effortless multiplicity. It follows that even the most fundamental ideas are expressed by a complex system of complementarities between sounds and shapes, each with its own meaning.

Kata is a characteristic example of this. Fans of martial arts are familiar with the concept of kata (型—form), the stylised movement pattern. Art forms like the *kabuki* theatre and calligraphy, and traditional practices, like the tea ceremony, each have their own kata; they are precisely designed to the point of being totally choreographed and perfectly synchronised movements of the participants which have remained almost unchanged for centuries to both preserve and convey meanings. Kata delineates and coordinates relationships between the participants and their lineage, the process, and the objects they use in the process. It also defines the relationships between the performers and the audience, the environment, and the historical continuity of the process. In industry and business, kata (型) denotes a model, a mould, a design, and a stereotype. Kata (written with the character 方 which means both direction or side and person and is also a way to refer to someone in a respectful way) is used to express the meaning of 'how', as, for example, in the phrase 'I know how to use this machine', and mode. Finally, kata (written with the character 肩 meaning shoulder) is used in conjunction with other characters in verbs such as offer to lend support (肩を貸す- pronounced kata o kasu), support or side with someone (肩を持つ– katawomotsu), being on equal terms with someone (肩を並べる—kata o naraberu), and to actively take over someone else's weight, debt, or responsibility (肩代わり—katagawari) but also in a passive sense as in to feel relieved because someone else has taken a burden off your shoulders (肩の荷が下りる—katanonigaoriru).

This brief linguistic diversion serves a set of interrelated purposes. First, it demonstrates the links between meanings, sounds, symbols, and images. We can clearly see a person moving forward in 方, the stylisation of art forms in 型, and the robustness of a strong shoulder you can lean on in 肩. Second, there is a comfortable coexistence of two writing systems in one word (the kanji, such as 肩 and 代, and the hiragana わ and り in 肩代わり). In a sentence it is also common to use a third syllabary, katakana, to write words of foreign origin (Sato Habein, 1984), and, of course, in modern cityscapes and on local as well as imported products, the Latin alphabet, and some really entertaining English, are omnipresent and as much a part of the average Japanese person's communication instruments as the native ones. Third, and most importantly, it highlights the characteristic fluidity, multidimensionality, and multifacetedness of Japanese thought. To understand their stark contrast to the Western linearity, univocality, and monosemy, just type 'do, doing, and will do' in Google Translate™. To save you the trouble, what you get is shown in Fig. 2.5.

Further to challenging the determinism and universality of the belief in the clear-cut dichotomy between juxtapositions through the existence of one unequivocal single truth, we hope to have highlighted that means, procedures, people, and their relationships are not distinct and independent constructions and entities but a unity of purpose and meaning. Kata implies that the person and other people, objects, the environment, and the process are integral and inseparable components of a standardised way of seamlessly coordinating their purposeful coexistence. The process frames and situates the person in space and time and in relation to the other participants in the process and those observing it or using its products. Within the process, the person, who is respected, supported, relied upon, and responsible for the support of others through the process, finds meaning but also gives life

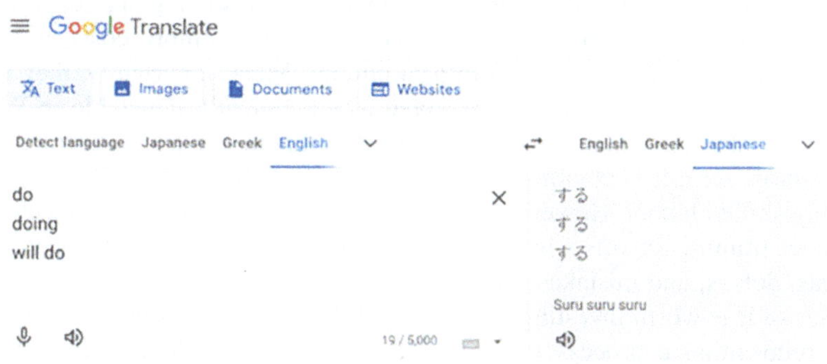

Fig. 2.5 Screenshot of the translation of the present and future forms of the verb 'to do' from English to Japanese in Google Translate™

and meaning to the process. Thus, adherence to the process is not an externally imposed 'must' but a central life choice for the individual, a definer of the group, and the métron of each and of their interrelationships.

Kata is a supportive mechanism of teamwork and a facilitator of a person's collective and perennial existence. In this way it alleviates existential angst; it fends off the fear of loneliness and death. Stylised as the process might be in its conception and design, every performance changes it, adds layers of meanings, and infuses it with subtle shades of character. An individual contribution might be imperceptible, but it contributes to keeping it alive through the ages, and so kata is morphed by its performers and observers while it morphs them into who they are and how they are connected with people, groups, nature, and objects in space and time.

A Case in Point: The Japanese Construction Industry

Kata applies to primary school children learning to write (you cannot find a word in a dictionary or a character on the computer keyboard if you do not know the precise order of strokes for each character) in exactly the same way as it applies to the works of the calligraphy masters. It is the way of doing even the most technical or mundane of things. A typical example of kata is construction, regardless of the building being a small house or a skyscraper. The Japanese construction industry is governed by the principles of mottainai and kata. These are expressed in the sustainable design, which is based on reconfiguring existing forms, and the standardisation of materials, building elements, and processes to the point of the industrial-scale prefabrication. The building elements are factory-produced and assembled on site in a quasi-ceremonial manner (Fraser & Zarkada-Fraser, 2000; Smith, 2009).

All construction sites in Japan are enclosed inside a protective cover, and they are kept devoutly neat and clean. The works do not cause noise and dust pollution and are designed to minimise any kind of nuisance for nearby residents and passers-by. Nevertheless, there are signs everywhere respectfully apologising for the inconvenience. Thanks to the consistent and painstaking application of the mottainai and kata principles, design, materials, and assembly costs as well as assembly labour-hours are reduced to a minimum. Most importantly, they reduce work accidents (which are largely accepted as endemically inevitable in Western societies, an obviously major moral and social problem, and a significant cost factor) to almost zero. It is obvious that such a construction site can only be operated by highly skilled labour. However, the relatively high costs of initial education and continuous training are offset by the significant cost savings in time, labour, wasted materials, delays, and mistakes. Moreover, in the spirit of mottainai, knowledge is reusable, so it is worth investing in it to avoid the risks and costs of having cheap labour reinventing a process for every edifice (Fraser & Zarkada-Fraser, 2000; Smith, 2009).

Mottanai and kata are the underlying principles of some of the most outstanding achievements of the Japanese construction industry. The world has repeatedly marvelled at the incredible speed with which Japan is able to heal the wounds of devastating natural disasters. For example, the 150 m long, and at places more than four metres deep, crater-like cracks left on the Great Kanto Highway in Naka by the Mw 9.0–9.1 undersea megathrust earthquake on 11 March 2011, which cost the lives of almost 20,000 people and caused the Fukushima disaster, were fully repaired in just six days (Mail Foreign Service, 2011). During the reconstruction, traffic was fully maintained. It is no wonder then that the most prestigious online construction information portal describes the Japanese construction industry as "an industry that executes complex projects with military precision and impeccable coordination", inconspicuously, quietly, and effectively "like a ninja" (Clegg, 2008).

Academic research has shown that "Japanese contractors achieve shorter construction times and higher levels of time certainty… anticipated delays are far shorter in Japan and levels of client satisfaction are significantly higher than in the US and UK (Xiao & Proverbs, 2002)". This yields reputational results and supports the industry's significant expansion overseas. Japanese companies, such as Sekisui House, are taking over notoriously hard-to-penetrate markets, such as housing in the US. The importance of having a higher purpose built into the fabric of an organisation and being guided by a focus on value and values is shown in this piece of news from the New York Post: "Sekisui House plans to send a team of 20 experts to the U.S. to begin training local contractors (Griffith, 2025)" not only to counter the labour shortages but, most importantly, to increase productivity and coordination, maintain the kata and safeguard mottainai. "Japanese construction workers are multidisciplinary", so "homes are typically built with far fewer workers" and "the average Japanese construction worker produces about 37% more new homes each year than the average U.S. construction worker" (Griffith, 2025). This, of course, produces significant financial results not only at the corporation but also at the industry level.

The financial consequences of consistently practicing kata and mottainai are that the Japanese construction industry is characterised by unusually high profitability and has been described as "super-efficient" (Clegg, 2008). In 2023, productivity in the Japanese construction sector reached approximately 104.8% (with 2020 being the benchmark year of 100%) while the U.S. and UK were measured at 85% and 90%, respectively. In the 1990s, average profitability exceeded 5%, while in 2008, despite the economic crisis and the contraction of government spending, profitability approached 3% (Crosthwaite & Connaughton, 2009). Average profitability in 2023 exceeded 4.5%, outperforming many Western counterparts (e.g. 3% in the U.S. and 2.8% in the UK) (Chen et al., 2023; Fraser & Zarkada-Fraser, 2000; Japan Build, 2024; ResearchAndMarkets.com, 2024; TechSci Research, 2025; Xiao & Proverbs, 2002; Zhang-Zhang & Kikkawa, 2023).

A significant part of the profits is invested in workforce education and development and in research on conventional environments, such as the development of new materials, efficient and safe processes, and research into the social impact of construction (Fraser & Zarkada-Fraser, 2000). Joint research between industry and universities is also underway on problem-solving solutions such as earthquake shock absorption systems using huge water tanks on the rooftops of high-rise buildings and futuristic projects such as the design of buildings that will stand one kilometre high, underwater and lunar cities, cities that grow like plants or float above the natural landscape, and lunar hotel complexes.

Time fluidity, the past-present-future unity, and principles such as kintsugi are not lost in the fast-increasing automation of the industry which is incorporating AI and robotics into the kata. The tradition of the fully biodegradable home, which is built of local natural materials like clay, wood, cotton, and bamboo so as to leave no mark behind when it is gone, is anything but abandoned. Japan leads in sustainable construction practices, with over 65% of new projects incorporating eco-friendly technologies, way above the 50% in the US and 45% in the UK, and notwithstanding the respective degrees of urbanisation or large-scale high-rise construction (Japan Build, 2024; ResearchAndMarkets.com, 2024; TechSci Research, 2025). Environmental sustainability goes hand in hand with technological innovation, thus yielding smart green construction practices. Builders are embracing eco-friendly methods such as energy-efficient designs, recycled materials, and renewable energy integration for the full building life cycle. Smart technologies, including IoT devices, sensors, and automation systems, are revolutionising building management, enabling energy optimisation, enhanced security, and maintenance efficiency (TechSci Research, 2025). This forward-thinking approach, combining environmental stewardship with cutting-edge technology, is also evident in the growing demand for renovation and remodelling. The following passage (Japan Build, 2024) aptly represents the Japanese way of "making the best possible use of whatever there is available in an organised and cooperative way".

> Younger generations, driven by financial constraints and a desire for modern living spaces, are increasingly opting to renovate old homes as a cost-effective alternative due to the demand for new condos and apartments outpacing supply. The Bank of Japan's ultra-low interest rate policy has made homeownership more viable, driving second-hand home sales. The Japanese Ministry of Land, Infrastructure, Transport and Tourism (MLIT) is addressing vacant houses by promoting renovations, repurposing homes as second residences or share houses, boosting renovation demand, and revitalizing local economies.

Having fully espoused the goal to become carbon neutral by 2050, Japan is using its construction industry as the spearhead of addressing geopolitical and sociodemographic challenges in its own way: kata, mottainai, kintsugi, and above all *ganbatte!* (頑張って—translated as 'Do your best!' or 'Keep going!' but also as 'I will give it my all')—the ethos of perseverance, determination, and effort, a collective mindset that values hard work, resilience, and a positive attitude in the face of challenges.

2.6 The Benefits of Finding and Sustaining Organisational Métron

Organisations which consistently serve a higher purpose find their métron in the clarity of their reason for their being, which unites their people. Consistently staying within their métron gives them continuity and allows them to enjoy the committed loyalty of their people. When their wholehearted commitment to serving their higher purpose is clearly communicated to external stakeholders and society at large, through products, acts, and narratives, organisations find their socially constructed and widely accepted métron of their existence. The more consistent and transparent they are in their serving their higher purpose, the more sustainable their métron and the closer they get to earn the Holy Grail of organisational performance: legitimacy.

Legitimacy is a generalised notion that an entity's actions are desirable, acceptable, or appropriate within a socially structured system of rules, values, beliefs, and definitions (Suchman, 1995). Legitimacy is a widely held perception. This means that it is socially constructed, and therefore objective. It is, however, based on interpretations of the historical continuity and consistency of the organisation's behaviour and evaluations of its people, actions, and products. In this sense it is subjectively constituted by each one of the organisation's stakeholders. It is crystallised into organisational reputation (Puncheva, 2008), defined as the understanding of the organisation as having an independent ontological status which stakeholders can share, experience, evaluate, appreciate, and influence (Zarkada & Polydorou, 2014). The actions of the organisation are publicised, they are commented on, and repeated, thus creating impressions and building perceptions which crystallise in the reputation of the organisation. Some researchers consider reputation as an asset of an organisation (Fombrun, 2001) and some as a resource (Goldberg et al., 2003) but everyone agrees that it is ubiquitous, important, and central to organisational (Zarkada & Polydorou, 2014). People love gossip, and reputation is the result of many people's gossiping over a long period of time and multiple media. If actions and marketing communications do not carefully structure and maintain it systematically, then people and social groups will construct it by gleaning fragmentary information from sources that may be aiming to erode the legitimacy of the organisation.

A base level of legitimacy ensures passive acceptance of an organisation as a member of the society which hosts it. High levels of legitimacy enlist the active support of stakeholders who perceive the organisation as remarkable, predictable, trustworthy, and meaningful. Legitimacy takes three distinct forms which reflect the interplay of perceptual and behavioural dynamics (Suchman, 1995): pragmatic, cognitive, and moral legitimacy.

2.6.1 Pragmatic Legitimacy

Pragmatic legitimacy refers to the calculations of the benefits that each of the interested parties expects from their relationship with the organisation. It is the result of communication of information and data, i.e. rational appeals, aimed at transforming an amorphous and indifferent audience into positively disposed stakeholders. Pragmatic legitimacy is the necessary basis of relationships between an organisation and the community in which it is hosted (Puncheva, 2008). Thus, without it, the organisation cannot accomplish its purpose. It is a three-dimensional construct, comprising exchange, influence, and dispositional legitimacy (Suchman, 1995).

Exchange legitimacy manifests as the willingness to enter into exchange relationships with the organisation as customers, suppliers, employees, supporters, etc. It is expressed in the value of the corporate brand and is directly related to the achievement of the organisation's goals. The more people believe that an organisation performs its function, and the more they expect it to continue to do so in the future, the higher the level of exchange pragmatic legitimacy the organisation enjoys (Suchman, 1995).

Similar to exchange legitimacy, but more socially than individually manifested, is influence legitimacy which makes stakeholders believe that the organisation serves their wider interests even if they do not intend to have direct dealings with it. Society, for example, supports the existence of universities where they will never work or study, museums they do not intend to visit, and even political parties with which they completely disagree. Measure of this form of legitimacy is the number of followers an organisation has on social media and the levels of positive engagement with its posts (Georgoudi & Zarkada, 2015; Zarkada, 2021; Zarkada & Polydorou, 2014).

In recent decades, organisations have been increasingly humanised and treated as autonomous, cohesive and ethically accountable entities (Suchman, 1995). However sociologically paradoxical it may be, there is often a generalised perception of organisations in terms of human character (good, reliable, prudent, ethical, etc.). This form of pragmatic legitimation is dispositional legitimacy (Suchman, 1995). A lot of research has been carried out on the purposefully generated anthropomorphism of brands through narratives, which act as cultural productions and vehicles for mythologising the brand and assigning meanings to it (Belk & Kniazeva, 2018; Kniazeva & Belk, 2010; Mills & John, 2021). Although there is relatively less academic interest on corporate brand management, a lot of what we know about product brand anthropomorphism applies to organisational brand storytelling and its role in shaping attitudes towards organisations at the individual and collective levels (Spear & Roper, 2013; Zimand-Sheiner, 2024). Marketing communications aim to generate dispositional legitimacy through CSR actions, stories of caring for employees, participation in environmental protection initiatives or initiating resource conservation practices, etc. Developing dispositional pragmatic legitimacy is also the main objective of rational appeals in Public Relations (PR), especially when trying to reverse negative attitudes towards the organisation caused by crises (e.g. workplace accidents, leaks of harmful waste into the environment, defective products, etc.).

2.6.2 Cognitive Legitimacy

Cognitive legitimation is about understanding and accepting the organisation, regardless of the personal interest or value judgment of the public. In a chaotic social environment, the identification of an organisation with accepted cultural norms makes it comprehensible and therefore acceptable. The more the organisation is understood as fitting in society, the higher the levels of cognitive legitimacy it enjoys. This results in its achieving the status of a stable intersubjective reference point for people which brings order to the chaos of subjective experience (Suchman, 1995).

Demonstrating that organisational practices adhere to accepted cultural norms is often restricted to paying lip service to popular ideals and utilising perception management techniques to make the organisation appear acceptable without really changing its characteristics that place it outside the boundaries of social acceptability. For example, the well-known fast-food chain Kentucky Fried Chicken™ employed the symbolic behaviour of changing its name to KFC™ to regain cognitive legitimacy following a wave of negative publicity about the unacceptable poultry farming practices implemented by its suppliers and the health risks associated with fried food consumption (Elsbach, 2014). With a drastic, but essentially immoral, intervention in the content of its marketing communications, the company distanced itself from outdated cultural paradigms without resorting to costly but positive alternatives such as product improvement and supply chain redesign.

Moreover, cognitive legitimation can be pursued through associating the organisation with institutions and organisations that are taken for granted. Believers, regardless of religion and creed, as well as atheists, take the Catholic Church, for example, for granted. In principle, an association with the Catholic Church makes the organisation also taken for granted. The association might be direct and pronounced, as in joining the UN Global Compact and proudly displaying its logos on digital and print corporate media. It can also be demonstrated through marketing communications connecting the organisation with generally accepted and respected institutions through symbolic behaviours (such as donating to Médecins sans Frontières) and physical markers (such as displaying a royal warrant on product packaging). Acts and markers can act as triggers or enhancements of cognitive legitimation. Consistently and systematically practicing what the organisation preaches is, however, a *sine qua non* for the achievement of a stable and effective cognitive legitimation.

2.6.3 Moral Legitimacy

Moral legitimacy manifests as positive value judgments of an organisation and its actions. These value judgments, like any moral evaluation, concern the intentions, means, ends, and virtue of the actors. It is based on information about the proclaimed purpose of the organisation, the procedures the organisation employs to fulfil its purpose, the results of the organisation's actions, and the character of the individuals who represent it (Suchman, 1995). Part of it is based on rumours and media coverage. Another part of it is the result of reputation-building marketing communications. This understanding has deep roots in the scholarly marketing tradition, even if it has been seen from a purely egoistical point of view. In his presidential address of the 1936 annual convention of the National Association of Marketing Teachers, Harvard Professor Harry R. Tosdal noted (Tosdal, 1936):

> Every manufacturing enterprise is likely to appear in the public eye, first as a smaller or larger capitalistic organization, that is, a corporate citizen of more or less selfish interest as a producer of profit; second, as an employer of a body of men and women; and, third, as a producer, buyer, and seller of commodities. If in its capacity as producer or vendor it is to achieve low costs of distribution, it is not sufficient to produce a good product and sell it at a market price. It is necessary also in these days more and more to have a reputation as a good employer, as a desirable citizen, so as to be relieved of added costs of governmental regulation and restrictions.

The content of marketing communications aimed at achieving the moral legitimacy of the organisation is emotionally oriented, includes symbolic acts and displays of altruism, and highlights a broader social role of the organisation that counterbalances perceptions of speculation or cynicism. The logic of the content of these communications is that of proclaiming legitimate authority as described by Max Weber (Weber, 1978/[1922]). Moral legitimacy, therefore, is the necessary foundation of society allowing the organisation to exist within its framework. Thus a modicum of moral legitimacy translates as societal consensus towards hosting the organisation in its midst (Puncheva-Michelotti et al., 2014). Without it, stakeholders do not recognise the organisation's right to operate and so they may turn against it, demand the removal of its facilities from their area, boycott its products, and even sabotage its property (Puncheva-Michelotti et al., 2010; Zarkada-Fraser & Fraser, 2002). High levels of moral legitimacy are the result of an organisation serving a higher purpose which is consistent with societal values and pursuits. People perceive the organisation as 'one of us', an entity working towards the common goals and contributing to meeting the objectives they have set for themselves, and, most importantly, as integral to building a world as they envisage it.

Moral legitimacy was largely overlooked in the framework of post modernity in which traditional values and the sense of community were ripped apart. The five new conditions of the culture of the late twentieth century, namely the engagement with hyperreality, fragmentation, reversed production and consumption, decentred subjects, juxtaposition of opposites and tolerance—if not fixation with cultivation and celebration—of difference and multiplicity (Anderson, 1995; Baudrillard, 1970/1988; Firat & Dholakia, 2006; Firat et al., 1995; Firat & Venkatesh, 1993) have played a key role in the development of postmodern marketing. The fruits of the labours of postmodern marketing are anthropomorphic brands as simulacra of the 'personality', 'identity' and 'image' of corporations (Kitchen & Proctor, 2015; Proctor & Kitchen, 2002) as well as the bearers of meanings strong enough to enmesh the most mundane of products into the life-fabric of consumers (Belk & Kniazeva, 2018; Kniazeva & Belk, 2010). Just like all simulacra, however, abstractions such as brands inescapably take the "path of all images: from a reflection of basic reality, to masking and perverting basic reality, to masking the absence of reality, to finally bearing no relation to any reality whatever, thus becoming its own pure simulacrum (Baudrillard, 1981 *trans* 1994)". They can mask the absence of a higher purpose for a short time, maybe in times of abundance and overproduction, but they are fast becoming irrelevant in times of shortages, recurring market failures, increasing poverty, and shrinking household disposable incomes. The disillusioned Boomers and Gen Xers, together with the ethical generations of their children and grandchildren, who are the mass of contemporary consumers and workforce, increasingly turn to the pursuit of substance over imagery, for at least a decade now (Panigyrakis & Zarkada, 2014a). Thus, brands are facing the fate of the "vast Map" in the Jorge Luis Borges *On Exactitude in Science* story (Borges, 1998): people "saw that that vast Map was Useless, and not without some Pitilessness was it, that they delivered it up to the Inclemencies of Sun and Winters".

Moreover, what has been just criticism of marketing practices by academics (Panigyrakis & Zarkada, 2014b) is fast materialising into public awareness, changing consumption patterns, and a drive towards conservatism, a search for meaning in traditional values, and a return to communitarianism in the form of rising nationalism and increase in ethnic and religious self-determination drives. People are losing patience with business practices "aiding and abetting of greed, individualism and hedonic consumption, if not.. bearing full responsibility for the rampant consumerism and subsequent non-sustainable business practices of the late twentieth and early twenty-first centuries (Panigyrakis & Zarkada, 2014b)".

Organisations serving a higher purpose can cut through the clutter of the fragmented, decentred, and juxtaposed subjectivities of postmodernity by refocusing on production and distribution of fairly priced products and services of value instead of promoting the consumption of egotistical and whimsical fantasies irrespective of their social and environmental consequences. Thus, embracing a higher purpose, employing value-dominant logic (Lindhult et al., 2018; Mahajan et al., 2023), and applying métron in their intentions, means, and ends, future managers can free their organisations from the straightjacket of amoral consumerism. Organisations led by future managers who practice axiological societing, that is, they hold themselves accountable and assume full responsibility for the social and environmental consequences of the choices they make on the basis of fundamentally axiological judgements (Panigyrakis & Zarkada, 2014a), are the ones who achieve high levels of moral legitimacy. Thus, they not only become focal points for the increasing numbers of people who reject "their alienation into decentred individualities through the erosion of communal values and overemphasis on the short-term quenching of desires (Panigyrakis & Zarkada, 2014b)" but they also act as catalyst of change, ushering an age of sustainable prosperity for all.

Future Management in Practice: Sustainable Prosperity
This chapter's focus has been on providing a *Purpose Redefinition.* As sustainable prosperity reshapes the goals of a manager, it represents the challenges of measuring prosperity through growth or material gains. It positions managers as custodians of balance between the present and the future, between the individual and the collective, and between the human and ecological systems.

1. From Growth to Stewardship

Future Management poses a challenge to managers to rethink prosperity, not as the endless expansion of markets, but as the stewardship of resources, relationships, and reputations. It includes designing business models that regenerate rather than deplete, and evaluating success by the durability of value created for all stakeholders rather than quarterly profit.

2. Intergenerational Thinking

Managers must cease focusing on immediate returns and extend their planning horizon. Future returns should be assessed on decisions: 'Will this product, process, or investment strengthen the organisation's legitimacy ten or twenty years from now?' Future management makes intergenerational equity a managerial and not an abstract societal goal.

3. Embedding Universality

Prosperity is always collective in nature. For sustainable prosperity, managers must broaden the definition of stakeholders beyond immediate groups, to include future generations, non-human species, and vulnerable populations that traditional accounting ignores, along with immediate groups, drawing from the accountability perspective discussed in Chapter 1.

4. Operational Implications

To truly achieve sustainable prosperity, managers must integrate environmental and social impact metrics into their core performance dashboards. They need to build supply chains with resilience and fairness in mind rather than focusing solely on cost efficiency. They must also rethink products and services to ensure circularity, reducing waste, and enhancing reuse. Managers should likewise participate in and fund knowledge-sharing initiatives and partnerships that enable collective problem-solving, understanding that prosperity must be sustained through collaboration rather than through competition.

5. Managerial Mindset Shift

The core of Future Management in this area is the regeneration of mindset. The role of managers transcends the arrangement of resources to become custodians of métron. There should be an equilibrium between surplus and shortfall, between individual profit and the public benefit. It calls for nurturing humility, a long-term outlook, and the readiness to embrace choices that forfeit short-term benefits in favour of enduring prosperity.

In practice:

Future Management in this context makes sustainable prosperity the yardstick of organisational legitimacy. Managers who accept this transition will transcend mere existence in a world rife with permanent crises and will establish organisations that endure, add value, and inspire for many generations to come.

References

Aaker, D. A. (2020). Winning against a dominant brand. *Journal of Brand Strategy, 9*(2), 103–112.

Acemoglu, D., & Robinson, J. A. (2012). *Why nations fail: The origins of power, prosperity, and poverty*. Profile Books.

Adhya, R. (2024). Mottainai: A holistic value stream approach to create sustainability. *Journal of Supply Chain Management, Logistics and Procurement, 6*(4), 383–396.

Akubra. (2024). *Australia's most iconic hat*. https://akubra.com.au/

Anderson, B. (1991). *Imagined communities: Reflections on the origins and spread of nationalism*. Verso.

Anderson, W. T. (1995). *The truth about the truth: De-confusing and re-constructing the postmodern world*. Jeremy Tarcher/Putnam.

Aoki, M. (1988). *Information, incentives and bargaining in the Japanese economy*. Cambridge University Press.

Aristotle. (*trans.* 1954/[4th century BCE]). *The Nicomachean ethics* (D. Ross, Trans.). Oxford University Press.

Aristotle. (*trans.* 2004/[4th century BCE]). *Rhetoric (Ρητορική)* (W. R. Roberts, Trans.). Kessinger Publishing.

Arnould, E. J., & Thompson, C. J. (2005). Consumer culture theory (CCT): Twenty years of research. *Journal of Consumer Research, 31*(4), 868–882.

Ashna, M. U., & Bandyopadhyay, S. (2024). Multi-objective assessment for industrial sustainability. *Clean Technologies and Environmental Policy, 26*, 2075–2076. https://doi.org/10.1007/s10098-024-02917-9

Atkinson, A. A., Waterhouse, J. H., & Wells, R. B. (1997). A stakeholder approach to strategic performance measurement. *MIT Sloan Management Review*.

Backstrom, L. (2011). Anatomy of Facebook. *Facebook Data*. https://www.facebook.com/notes/facebook-data-team/anatomy-of-facebook/10150388519243859

Barry, E. (2016, 04/09/2016). Dalai Lama offers musings to students: Mosquitoes are ungrateful, for starters. *New York Times*. https://www.nytimes.com/2016/04/09/world/asia/musings-from-the-dalai-lama-mosquitoes-are-ungrateful-for-starters.html

Battelle, J. (2005). *The search: How google and its rivals rewrote the rules of business and transformed our culture*. Nicholas Brealey.

Baudrillard, J. (1981 *trans* 1994). *Simulacra and Simulation (Simulacres et simulation)* (S. F. Glasser, Trans.). The University of Michigan Press.

Baudrillard, J. (1988). *The consumer society: Myths and structures (Société de Consommation: Ses Mythes, Ses Structures)* (C. Turner, Trans.). Sage. (Original work published 1970)

Bauza, V. (2023). Methuselah: Still the world's oldest tree? *Conservation International*. https://www.conservation.org/blog/methuselah-still-the-worlds-oldest-tree#:~:text=In%20eastern%20California%2C%20a%20Great,built%20the%20pyramids%20at%20Giza.

Beekun, R. I., & Badawi, J. A. (2005). Balancing ethical responsibility among multiple organizational stakeholders: The Islamic perspective. *Journal of Business Ethics, 60*(2), 131–145.

Bejczy, I. P. (2011). Virtue and vice. In H. Lagerlund (Ed.), *Encyclopedia of medieval philosophy* (pp. 1365–1369). Springer Netherlands. https://doi.org/10.1007/978-1-4020-9729-4_515

Belk, R., & Kniazeva, M. (2018). Morphing anthropomorphism: An update. *Journal of Global Scholars of Marketing Science, 28*(3), 239–247. https://doi.org/10.1080/21639159.2018.1466659

Benedict, R. (1954). *The chrysanthemum and the sword: Patterns of Japanese Culture*. Charles E. Tuttle Co.

Bentele, G., & Seidenglanz, R. (2008). Trust and credibility—Prerequisites for communication management. In A. Zerfass, B. van Ruler, & K. Sriramesh (Eds.), *Public relations research: European and international perspectives and innovations* (pp. 48–62). Springer.

Blair, M. M. (1998). For whom should corporations be run?: An economic rationale for stakeholder management. *Long Range Planning, 31*(2), 195–200.

Borges, J. L. (1998). Museum, on exactitude in science. *Collected fictions*, 325–327.

Botsman, R., & Rogers, R. (2011). *What's mine is yours: How collaborative consumption is changing the way we live*. HarperCollins e-books.

Bovens, L. (2020, September 4). 'From each according to ability; to each according to need'—Tracing the biblical roots of socialism's enduring slogan. *The Conversation*. https://theconversation.com/from-each-according-to-ability-to-each-according-to-need-tracing-the-biblical-roots-of-socialisms-enduring-slogan-138365

Bray, J., Johns, N., & Kilburn, D. (2011). An exploratory study into the factors impeding ethical consumption. *Journal of Business Ethics, 98*(4), 597–608. https://doi.org/10.1007/s10551-010-0640-9

Brundtland, G. H. (1994). What is world prosperity? *Business Strategy Review, 5*(2), 57–69.

Burns, D. J. (2015). What comes after postmodernism? Implications for marketers. *Journal of Global Scholars of Marketing Science, 25*(1), 59–74. https://doi.org/10.1080/21639159.2014.980035

Callanan, G. A. (2015). They reap but do not sow: How multinational corporations are putting an end to virtuous capitalism. *Business and Society Review, 120*(3), 363–384. https://doi.org/10.1111/basr.12059

Carlson, S. C. (2024). *Golden ratio*. In Encyclopædia Britannica.

Carroll, A. B. (1991). The pyramid of corporate social responsibility: Toward the Moral management of organizational stakeholders. *Business Horizons, July–August*(39–48).

Carroll, A. B. (1993). *Business & society: Ethics and stakeholder management* (2nd ed.). South-Western Publishing.

Carroll, A. B. (1998). The four faces of corporate citizenship. *Business and Society Review*(100/101), 1–7.

Carroll, A. B., & Buchholtz, A. K. (1996). Ethics and stakeholder management. *Cincinnati: South-Western.*

Catoméris, B. P. (2012). *Foundations of Yoga: The traditional teachings of Sri Shyam Sundar Goswami*. Inner Traditions.

Chatzopoulou, E., & de Kiewiet, A. (2021). Millennials' evaluation of corporate social responsibility: The wants and needs of the largest and most ethical generation. *Journal of Consumer Behaviour, 20*(3), 521–534.

Chen, W. T., Kiani, A. K., Wu, M.-T., Merrett, H. C., & Wang, C.-H. (2023). Comparison of the technical efficiency of construction industries—A case study of Taiwan and Mainland China. *Sustainability, 15*(2), 941.

Chen, Y.-R. R., & Zhao, X. (2021). Digital dialogue in online brand communities: Examining the social network outcomes of brands' dialogue with Facebook users. *Telematics and Informatics, 57*, 101507. https://doi.org/10.1016/j.tele.2020.101507

Cheng, Y. (2018). How social media is changing crisis communication strategies: Evidence from the updated literature. *Journal of Contingencies and Crisis Management, 26*(1), 58–68. https://doi.org/10.1111/1468-5973.12130

Clegg, J. (2008). Japanese construction: the hidden fortress. *Building*(15). Retrieved 07/09/2009, from http://www.building.co.uk/story.asp?storycode=3111274

Coffin, J., & Egan-Wyer, C. (2022). The ethical consumption cap and mean market morality. *Marketing Theory, 22*(1), 105–123.

Colish, M. L. (1990). *The stoic tradition from antiquity to the early middle ages: Stoicism in classical Latin Literature. I* (Vol. 1). Brill.

CompaniesMarketcap.com. (2024a, 20/08/2024). *End of Day market cap according to different sources.* Retrieved 21/08/2024 from https://companiesmarketcap.com/alphabet-google/marketcap/

CompaniesMarketcap.com. (2024b, 21/08/2024). *Market capitalization of Alphabet (Google) (GOOG).* Retrieved 21/08/2024 from https://companiesmarketcap.com/alphabet-google/marketcap/

Cooke, B. D. (2014). *Australia's war against rabbits: the story of rabbit haemorrhagic disease.* CSIRO publishing.

Covey, S. R. (2020). *The 7 habits of highly effective people: Powerful lessons in personal change* (30th Anniversary ed.). Simon and Schuster.

Crosthwaite, D., & Connaughton, J. (2009). The world construction outlook. *Building*(27/02/2009). Retrieved 07/09/2009, from http://www.building.co.uk/story.asp?sectioncode=661&storycode=3134710

Dale, P. N. (1990). *The myth of Japanese uniqueness.* Routledge.

Dannreuther, C. (2024). Power in the future of work: Production, reproduction, and reconstruction. *Review of Evolutionary Political Economy, 5*(2), 329–350.

Dwivedi, Y. K., Ismagilova, E., Hughes, D. L., Carlson, J., Filieri, R., Jacobson, J., Jain, V., Karjaluoto, H., Kefi, H., Krishen, A. S., Kumar, V., Rahman, M. M., Raman, R., Rauschnabel, P. A., Rowley, J., Salo, J., Tran, G. A., & Wang, Y. (2021). Setting the future of digital and social media marketing research: Perspectives and research propositions. *International Journal of Information Management, 59*, 102168. https://doi.org/10.1016/j.ijinfomgt.2020.102168

Eco, U. (Ed.). (2004). *Ιστορία της Ομορφιάς (History of Beauty).* Καστανιώτης (Kastaniotis).

Elsbach, K. D. (2014). *Organizational perception management.* Routledge.

Engels, F. (1998 & 2001/[1925]). *Dialectics of nature* (S. Ryan & jjazz@hwcn.org, Eds.). Marxists Internet Archive.

Ewing, J. (2017). Inside VW's campaign of trickery. *New York Times.* Retrieved 05/07/2017 May 07, from https://www.nytimes.com/2017/05/06/business/inside-vws-campaign-of-trickery.html

Fader, P. S., Hardie, B. G., & Lee, K. L. (2005). RFM and CLV: Using iso-value curves for customer base analysis. *Journal of Marketing Research, 42*(4), 415–430.

Fana, M., & Villani, D. (2024). A contemporary perspective on social classes and functional income distribution. *Review of Political Economy*, 1–26.

Firat, A. F., & Dholakia, N. (2006). Theoretical and philosophical implications of postmodern debates: Some challenges to modern marketing. *Marketing Theory, 6*(2), 123–162.

Firat, A. F., Dholakia, N., & Venkatesh, A. (1995). Marketing in a postmodern world. *European Journal of Marketing.* https://doi.org/10.1108/03090569510075334

Firat, A. F., & Venkatesh, A. (1993). Postmodernity: The age of marketing. *International Journal of Research in Marketing, 10*(3), 227–249. https://doi.org/10.1016/0167-8116(93)90009-N

FitzGerald, J., & McArthur, T. (2025). What's the latest on the Los Angeles wildfires and what caused them? *BBC News.* https://www.bbc.com/news/articles/clyxypryrnko

Fitzpatrick, A. (2014). A judge ordered Microsoft to split. Here's why it's still a single company. *Time.* Retrieved 30/12/2024, from https://time.com/3553242/microsoft-monopoly/

Flurry, L. A., Swimberghe, K. R., & Parker, J. M. (2014). Examining brand communities among children and adolescents: An exploratory study. *Journal of Consumer Marketing, 31*(2), 103–110.

Fombrun, C. J. (2001). Corporate reputation as economic assets. In M. A. Hitt, E. R. Freeman, & J. S. Harrison (Eds.), *The Blackwell handbook of strategic management* (pp. 289–312). Blackwell Publishers.

Ford, D. (Ed.). (1990). *Understanding business markets: Interaction, relationships and networks.* Academic Press.

Fraser, C., & Zarkada-Fraser, A. K. (2000, September 6–8). Research and development in the Japanese construction industry. Transcending Boundaries: Integrating People, Processes and Systems, Brisbane.

Fraser, C., & Zarkada-Fraser, A. K. (2001). The philosophy, structure and objectives of research and development in Japan. *Construction Management and Economics, 19*(8), 831–840.

Freeman, R. E. (2010/[1984]). *Strategic management: A stakeholder approach.* Cambridge University Press.

Friedman, A. L., & Miles, S. (2006). *Stakeholders: Theory and practice.* Oxford University Press. https://doi.org/10.1093/oso/9780199269860.001.0001

Friedman, M. (1970). The social responsibility of business is to increase its profits. *The New York Times Magazine, Sept. 13*, 32–33, 122–126. https://www.nytimes.com/1970/09/13/archives/a-friedman-doctrine-the-social-responsibility-of-business-is-to.html

Friedman, M. (1982). *Capitalism and freedom* (Reissue with new Preface by the Author ed.). The University of Chicago Press.

Fuchs, C. (2020). *Communication and capitalism: A critical theory.* University of Westminster Press. https://doi.org/10.16997/book45

Fuduric, M., & Mandelli, A. (2014). Communicating social media policies: Evaluation of current practices. *Journal of Communication Management, 18*(2), 158–175.

Galbraith, J. K. (1998/[1958]). *The affluent society* (40th annivesary ed.). Houghton Mifflin Harcourt.

Geddes, P. (1915). *Cities in evolution: An introduction to the town planning movement and to the study of civics.* Williams & Norgate.

Georgoudi, F., & Zarkada, A. K. (2015, May 26–29). *The application of the elaboration likelihood model to political communication: Political slogans and change of preference* 44th EMAC Annual Conference.

Goldberg, A. I., Cohen, G., & Fiegenbaum, A. (2003). Reputation building: Small business strategies for successful venture development. *Journal of Small Business Management, 41*(2), 168–187.

Griffith, K. (2025). Japanese company Sekisui House quietly becomes America's fifth-largest homebuilder. *New York Post.* Retrieved 24/01/2025, from https://nypost.com/2025/01/03/real-estate/japanese-company-quietly-becomes-americas-fifth-largest-homebuilder/#!

Gurova, O. (2017). Consumer culture in socialist Russia. In O. Kravets, P. Maclaran, S. Miles, & A. Venkatesh (Eds.), *The SAGE handbook of consumer culture* (pp. 102–104). Sage.

Hahn, R. (2022). *Sustainability management: Global perspectives on concepts, instruments, and stakeholders.* Rüdiger Hahn.

Hanley, S. B. (1999). *Everyday things in premodern Japan: The hidden legacy of material culture.* University of California Press.

Hardacre, H. (2025). *Are the Japanese people religious?* Japan Society. Retrieved 23/01/2025 from https://aboutjapan.japansociety.org/are_the_japanese_people_religious#sthash.RIeY8vN0.dpbs

Healy, J. C., & McDonagh, P. (2013). Consumer roles in brand culture and value co-creation in virtual communities. *Journal of Business Research, 66*(9), 1528–1540.

Helliwell, J. F., Layard, R., & Sachs, J. D. (2019). *World Happiness Report 2019.* Sustainable development solutions network. Retrieved 12/01/2025 from https://worldhappiness.report/ed/2019/#read

Helliwell, J. F., Layard, R., Sachs, J. D., De Neve, J.-E., Aknin, L. B., & Wang, S. (2023). *World happiness report 2023.* Sustainable Development Solutions Network. Retrieved 12/01/2025 from https://happiness-report.s3.amazonaws.com/2023/WHR+23.pdf

Hess, H. W. (1935). The new consumption era. *American Marketing Journal, 2*(1), 16–25.

His Holiness the 14th Dalai Lama. (2021). The purpose of life is to be happy. *India Today.* https://www.dalailama.com/messages/transcripts-and-interviews/the-purpose-of-life-is-to-be-happy

Holden, N. J., & Burgess, M. (1994). *Japanese-Led Companies: Understanding how to make them your customers.* McGraw-Hill.

Holy Bible. (2011). (New International Version ed.). Biblica, Inc.

Houssiau, F., Sapieżyński, P., Radaelli, L., Shmueli, E., & de Montjoye, Y.-A. (2023). Detrimental network effects in privacy: A graph-theoretic model for node-based intrusions. *Patterns, 4*(1), 1–12.

Hsu, T., & Myers, S. L. (2024). Europe banned Russia's RT network. Its content is still spreading. *The New York Times*. https://www.nytimes.com/2024/05/30/business/media/russia-rt-disinformation-europe-ban.html

Invasive Species Council. (2022). Cats in Australia. *Our Work*. https://invasives.org.au/our-work/feral-animals/cats-in-australia/

Istituto Nazionale di Statistica. (2024, 06/06/2024). *Italy's Economic outlook 2024–2025*. Retrieved 21/08/2024 from https://www.istat.it/en/press-release/italys-economic-outlook-2024-2025/

Iwabuchi, K. (2002). Recentering globalization: Popular culture and Japanese transnationalism. *Duke UP*.

Jackson, T. (2015). Preface. In I. Cassiers (Ed.), *Redefining prosperity*. Routledge London. https://doi.org/10.4324/9781315777955

Japan Build. (2024). *The Japan construction market: A comprehensive overview*.

Johnson, M. D., & Auli, S. (1998). Customer satisfaction, loyalty, and the trust environment. *Advances in Consumer Research, 25*(1), 15–20.

Johnson, R., & Cureton, A. (2024). Kant's moral philosophy. In E. N. Zalta & U. Nodelman (Eds.), *The Stanford encyclopedia of philosophy* (Fall ed.). Stanford University.

Jones, B. (2025). Why does Trump hate this tiny fish so much? As Los Angeles burns, President-elect Donald Trump is yelling at a fish. *Climate/Down to Earth*. Retrieved 12/01/2025, from https://www.vox.com/down-to-earth/394283/los-angeles-wildfires-trump-newsom-delta-smelt

Jones, T. M., Harrison, J. S., & Felps, W. (2018). How applying instrumental stakeholder theory can provide sustainable competitive advantage. *Academy of Management Review, 43*(3), 371–391.

Kamath, R. R., & Liker, J. K. (1994). A second look at Japanese product development. *Harvard Business Review, 72*(6), 154–165.

Kant, E. (1959/[1785]). *Groundwork for the metaphysics of morals* (L. W. Beck, Trans.; 1959 reprint ed.). Bobbs-Merrill.

Karinthy, F. (1929). Láncszemek (Chain-Links) (A. Makkai & E. Jankó, Trans.). In F. Karinthy (Ed.), *Everything is different*.

Kazantzakis, N. (2012/[1927]). *The saviors of god: Spiritual exercises* (K. Friar, Trans.). Simon & Schuster.

Kitchen, P. J., & Proctor, T. (2015). Marketing communications in a post-modern world. *Journal of Business Strategy, 36*(5), 34–42. https://doi.org/10.1108/JBS-06-2014-0070

Kleinfeld, J. (2002). The small world problem. *Society, 39*(2), 61–66.

Kniazeva, M., & Belk, R. W. (2010). If this brand were a person, or Anthropomorphism of brands through packaging stories. *Journal of Global Academy of Marketing Science, 20*(3), 231–238. https://doi.org/10.1080/12297119.2010.9707349

Knight, H., Haddoud, M. Y., & Megicks, P. (2022). Determinants of corporate sustainability message sharing on social media: A configuration approach. *Business Strategy and the Environment, 31*(2), 633–647. https://doi.org/10.1002/bse.2941

Knowles-Dellner, S. (2017). *Lagom*. Quadrille.

Kodama, F. (1995). *Emerging Patterns Of Innovaton: Sources of Japan's technological edge*. Harvard Business School Press.

Kravets, O., Maclaran, P., Miles, S., & Venkatesh, A. (2017). *The SAGE handbook of consumer culture*. Sage.

Kumar, P. (1999). The impact of long-term client relationships on the performance of business service firms. *Journal of Service Research, 2*(1), 4–18. https://doi.org/10.1177/109467059921002

Kumar, V., Venkatesan, R., Bohling, T., & Beckmann, D. (2008). Practice prize report—The power of CLV: Managing customer lifetime value at IBM. *Marketing Science, 27*(4), 585–599. https://doi.org/10.1287/mksc.1070.0319

Lahdili, N. (2015). Navigating environmental policies: A case study of Morocco's response to climate change. *Public Policy Analysis, PSCI, 6808*, 1–23.
Lazonick, W. (2009). *Sustainable prosperity in the new economy?: Business organization and high-tech employment in the United States.* WE Upjohn Institute.
Lee, K.-H. (2012). Linking stakeholders and corporate reputation towards corporate sustainability. *International Journal of Innovation and Sustainable Development, 6*(2), 219–235. https://doi.org/10.1504/IJISD.2012.046947
Lemon, K. N., & Verhoef, P. C. (2016). Understanding customer experience throughout the customer journey. *Journal of Marketing, 80*(6), 69–96.
Lenin, V. I. (1999/[1902]). *What is to be done?—Burning questions of our movement* (Vol. 5). Foreign Languages Publishing House & Marxist Internet Archive.
Li, F., Larimo, J., & Leonidou, L. C. (2021). Social media marketing strategy: Definition, conceptualization, taxonomy, validation, and future agenda. *Journal of the Academy of Marketing Science, 49*(1), 51–70. https://doi.org/10.1007/s11747-020-00733-3
Lindhult, E., Chirumalla, K., Oghazi, P., & Parida, V. (2018). Value logics for service innovation: Practice-driven implications for service-dominant logic. *Service Business, 12*(3), 457–481.
Littler, J. (2018). Celebrity CEOs and the cultural economy of tabloid intimacy. In S. Holmes & S. Redmond (Eds.), *Stardom and celebrity: A reader* (pp. 230–243). Sage. https://doi.org/10.4135/9781446269534.n22
Logsdon, J. M., & Yuthas, K. (1997). Corporate social performance, stakeholder orientation, and organizational moral development. *Journal of Business Ethics, 16*(12/13), 1213–1226. https://doi.org/10.1023/A:1005741931995
López-Fernández, A. M. (2020). Price sensitivity versus ethical consumption: A study of Millennial utilitarian consumer behavior. *Journal of Marketing Analytics, 8*(2), 57–68. https://doi.org/10.1057/s41270-020-00074-8
Lubbock, J. (2024). How pro-Israel groups shape global media coverage of Palestine. *Declassified UK*. Retrieved 12/01/2025, from https://www.declassifieduk.org/how-pro-israel-groups-shape-global-media-coverage-of-palestine/
Lubowiecki-Vikuk, A., Dąbrowska, A., & Machnik, A. (2021). Responsible consumer and lifestyle: Sustainability insights. *Sustainable Production and Consumption, 25*, 91–101. https://doi.org/10.1016/j.spc.2020.08.007
Luqman, A., Talwar, S., Masood, A., & Dhir, A. (2021). Does enterprise social media use promote employee creativity and well-being? *Journal of Business Research, 131*, 40–54. https://doi.org/10.1016/j.jbusres.2021.03.051
Mackie, J. L. (1990). *Ethics: Inventing right and wrong*. Penguin.
Mahajan, G., Kumar, V., Tregua, M., & Bruni, R. (2023). Value-dominant logic: Organizational principles. *European Journal of Marketing, 57*(5), 1245–1271. https://doi.org/10.1108/EJM-11-2022-0816
Mail Foreign Service. (2011). The Japanese road repaired SIX days after it was destroyed by quake. *Mail Online*. Retrieved 23/01/2025, from https://www.dailymail.co.uk/news/article-1369307/Japan-tsunami-earthquake-Road-repaired-SIX-days-destroyed.html
Markovitz, G. (2024, December 9). *Davos 2025: What to expect and who's coming?* World Economic Forum. Retrieved 22/01/2025 from https://www.weforum.org/stories/2024/12/davos-2025-whos-coming-and-what-to-expect/
Marx, K. (2018/[1867]). *The capital* (Vol. 1–3). Madison & Adams Press.
Marx, K., & Engels, F. (1996). *The Communist Manifesto* (S. f. t. o. G. E. o. Moore, revised by Friedrich Engels, Trans.). Orion Books. (Original work published 1888).
McCabe, D. (2024). 'Google is a monopolist,' Judge Rules in Landmark Antitrust Case. *The New York Times*. Retrieved 21/08/2024, from https://www.nytimes.com/2024/08/05/technology/google-antitrust-ruling.html.
McCarthy, D. J., Puffer, S. M., Dunlap, D. R., & Jaeger, A. M. (2012). A stakeholder approach to the ethicality of BRIC-firm managers' use of favors. *Journal of Business Ethics, 109*(1), 27–38. https://doi.org/10.1007/s10551-012-1377-4

McLuhan, M., & Powers, B. R. (1989). *The global village: Transformations in world life and media in the 21st century*. Oxford University Press.

Mehra, K., & Sarkar, N. (2024). Nanomaterials for the management of plastic waste: A way combating white pollution. *Nanotechnology for Environmental Engineering, 9*(4), 801–815. https://doi.org/10.1007/s41204-024-00396-3

Milgram, S. (1967). The small world problem. *Phychology Today, 1*, 61–67.

Mills, A. J., & John, J. (2021). Brand stories: Bringing narrative theory to brand management. *Journal of Strategic Marketing*, 1–19. https://doi.org/10.1080/0965254x.2020.1853201

Moore, H. L., Davies, M., Mintchev, N., & Woodcraft, S. (2021). *Prosperity in the twenty first century: Concepts, models and metrics*. Institute for Global Prosperity, UCL.

Morishita, A., & Tainaka, K.-i. (2019). Cooperation promotes the sustainability of companies: Lattice-gas model for a market. *Physica A: Statistical Mechanics and its Applications, 525*, 119–127. https://doi.org/10.1016/j.physa.2019.03.005

Morita, A., Reingold, E. M., & Shinomura, M. (1988). *Made in Japan: Akio Morita and Sony*. Signet

Nakane, C. (1973). *Japanese society*. Penguin.

National Geographic. (2025). Trash and overcrowding at the top of the world. *Education*. https://education.nationalgeographic.org/resource/trash-and-overcrowding-top-world/

National Geographic Society. (2022). Biodiversity. In *Resource library*. National Geographic.

Nguyen, T. K. G., Ozawa, T., & Fan, P. (2024). Sanpo-yoshi, top management personal values, and ESG performance. *Journal of Behavioral and Experimental Finance, 41*, 100903. https://doi.org/10.1016/j.jbef.2024.100903

Nightingale, K. (2020). Multi-brand retailer community model: How multi-brand retailers can create and manage strong brand communities. *Journal of Brand Strategy, 9*(3), 240–253.

Nurisso, G. C., & Prescott, E. S. (2020). Origins of too-big-to-fail policy in the United States. *Financial History Review, 27*(1), 1–15.

Nuryana, M., Wiryanto, W., & Aswin, A. (2024). Intervention strategy for sustainable waste management systems in developing countries using community of practice. *Global Journal of Research, 10*, 323. https://doi.org/10.22034/gjesm.2024.10.SI.20

Packard, V. (1960). *The waste makers*. Penguin Books.

Panigyrakis, G. G., & Zarkada, A. K. (2014a). New philosophical paradigms in marketing: From amoral consumerism to axiological societing. In L. Moutinho, E. Bigne, & A. K. Manrai (Eds.), *The Routledge companion on the future of marketing*. Routledge.

Panigyrakis, G. G., & Zarkada, A. K. (2014b). A philosophical investigation of the transition from Integrated Marketing Communications to metamodern meaning cocreation. *Journal of Global Scholars of Marketing Science, 24*(3), 262–278. https://doi.org/10.1080/21639159.2014.911494

Pauluzzo, R., & Mason, M. C. (2022). A multi-dimensional view of consumer value to explain socially-responsible consumer behavior: A fuzzy-set analysis of Generation Y's fast-fashion consumers. *Journal of Marketing Theory and Practice, 30*(2), 191–212. https://doi.org/10.1080/10696679.2021.1908146

Perrigot, R., Kacker, M., Basset, G., & Cliquet, G. (2012). Antecedents of early adoption and use of social media networks for stakeholder communications: Evidence from franchising. *Journal of Small Business Management, 50*(4), 539–565. https://doi.org/10.1111/j.1540-627X.2012.00366.x

Pitts, E. (2024). Gen Z takes on the American dream: How pandemic-era challenges shaped Gen Z's outlook on adulthood. *Deseret News*. https://www.deseret.com/utah/2024/03/17/generation-z-housing-college-tuition-american-dream/#:~:text=%E2%80%9CGen%20Z's%20way%20of%20working,aim%20for%20a%20stable%20future.

Pope Francis. (2017). Angelus, Matthew 16: 21–27 *Happiness*. Retrieved 11/01/2025, from https://www.popefrancishomilies.com/happiness

Proctor, T., & Kitchen, P. (2002). Communication in postmodern integrated marketing. *Corporate Communications: An International Journal, 7*(3), 144–154. https://doi.org/10.1108/13563280210436754

Puncheva-Michelotti, P., McColl, R., Vocino, A., & Michelotti, M. (2014). Corporate patriotism as a source of corporate reputation: A comparative multi-stakeholder approach. *Journal of Strategic Marketing,* 1–23. https://doi.org/10.1080/0965254x.2014.885989

Puncheva-Michelotti, P., Michelotti, M., & Gahan, P. (2010). The relationship between individuals' recognition of human rights and responses to socially responsible companies: Evidence from Russia and Bulgaria. *Journal of Business Ethics, 93*, 583–605. https://doi.org/10.1007/s10551-009-0242-6

Puncheva, P. Y. (2008). The role of corporate reputation in the stakeholder decision-making process. *Business & Society, 47*(3), 272–290. https://doi.org/10.1177/0007650306297946

Putnam, R. D. (2001). *Bowling alone: The collapse and revival of american community* (Revised and Updated ed.). Simon & Schuster.

Putnam, R. D., & Garrett, S. R. (2020). *The upswing: How America came together a century ago and how we can do it again.* Simon & Schuster.

Ramos, C. M. Q., & Casado-Molina, A.-M. (2021). Online corporate reputation: A panel data approach and a reputation index proposal applied to the banking sector. *Journal of Business Research, 122*, 121–130. https://doi.org/10.1016/j.jbusres.2020.08.061

Raval, A. (2024). The disinformation storm is now hitting companies harder. *Financial Times.* Retrieved 26/12/2024, from https://www.ft.com/content/0aa9725d-e423-4a6b-b842-866ad4541dc2

Ray, R. S., Hornibrook, J., Skitmore, M., & Zarkada-Fraser, A. K. (1999). Ethics in tendering: A survey of Australian opinion and practice. *Construction Management and Economics, 17*(2), 139–153.

Reading, B. (1993). *Japan: The coming collapse.* Orion Books.

Rear, D. (2017). A critical analysis of japanese identity discourse: Alternatives to the Hegemony of Nihonjinron. *Asian Studies: Journal of Critical Perspectives on Asia, 53*(2).

ResearchAndMarkets.com. (2024). *Japan Construction Industry Databook 2024—Market Size & Forecast by Value and Volume (area and units)* (Japan Construction Industry Databook Series -),, Issue. https://www.researchandmarkets.com/reports/5918174/japan-construction-industry-databook-series

Robertson, R. (1992). *Globalization: Social theory and global culture.* Sage Publications.

Rojek, C. (2001). *Celebrity.* Reaktion Books.

Rowley, J., & Keegan, B. (2020). An overview of systematic literature reviews in social media marketing. *Journal of Information Science, 46*(6), 725–738. https://doi.org/10.1177/0165551519866544

Russell, B. (1930). *The conquest of happiness* (Full Text ed.). Unwin Brothers Limited/Portal Site for Russellian in Japan.

Sadowski, J. (2020). The internet of landlords: Digital platforms and new mechanisms of rentier capitalism. *Antipode, 52*(2), 562–580. https://doi.org/10.1111/anti.12595

Sato Habein, Y. (1984). *The history of the Japanese written language.* University of Tokyo Press.

Schumpeter, J. A. (2013/[1942]). *Capitalism, socialism and democracy* (1st ed.). Routledge.

Scruton, R. (2012). *Modern philosophy: An introduction and survey.* A&C Black.

Seneca, L. A. t. Y. (2004). *Για μια Ευτυχισμένη Ζωή [De Vita Beata].* Πατάκης (Patakis). (Original work published 50 BCE).

Shinohara, Y., & Kim, R. C. (2022). Sanpo-Yoshi and corporate social responsibility in Japan. *Strategic Analysis, 46*(4), 403–415. https://doi.org/10.1080/09700161.2022.2111763

Singh, S. (2024, December 26). *How many people use social media (2025 Statistics).* DemandSage. Retrieved 13/01/2025 from https://www.demandsage.com/social-media-users/

Singh, S. S., & Jain, D. C. (2014). Evaluating customer relationships: Current and future challenges In L. Moutinho, E. Bigné, & A. K. Manrai (Eds.), *The Routledge companion to the future of marketing* (pp. 193–217). Routledge.

Skorkin, K. (2024). Ukraine's Ban on Moscow-Linked Church will have far-reaching consequences. *Carnegie Politica.* Retrieved 23/01/2025, from https://carnegieendowment.org/russia-eurasia/politika/2024/08/zapret-upc-v-ukraine?lang=en

Smalls, B. B. (2025). Facebook littered with fake headlines after Mark Zuckerberg announced meta will no longer be fact-checking. *HipHopWired*. https://hiphopwired.com/playlist/mark-zuckerberg-fake-facebook-headlines-joe-rogan-podcast-reactions/item/7/

Smith, R. E. (2009, May 20–24). *History of prefabrication: A cultural survey*. Proceedings of the Third International Congress on Construction History, Cottbus, Germany. http://www.tu-cottbus.de/bautechnikgeschichte/ch2009/openaccess/smith_oa.pdf

Sood, S. C., & Pattinson, H. M. (2012). 21st Century applicability of the interaction model: Does pervasiveness of social media in B2B marketing increase business dependency on the interaction model? *Journal of Customer Behaviour, 11*(2), 117–128. https://doi.org/10.1362/147539212X13420906144598

Spear, S., & Roper, S. (2013). Using corporate stories to build the corporate brand: An impression management perspective. *Journal of Product & Brand Management, 22*(7), 491–501. https://doi.org/10.1108/jpbm-09-2013-0387

Standing, G. (2021). *The corruption of capitalism: Why rentiers thrive and work does not pay*. Biteback Publishing.

Stiglitz, J. E. (2020). GDP is the wrong tool for measuring what matters. *Scientific American*. Retrieved 25/07/2022, from https://www.scientificamerican.com/article/gdp-is-the-wrong-tool-for-measuring-what-matters/

Suchman, M. C. (1995). Managing legitimacy: Strategic and institutional approaches. *Academy of Management Review, 20*(July), 571–610. https://doi.org/10.5465/amr.1995.9508080331

Tamashiro, T. (2019). *How to Ikigai*. Mango Media.

Tasker, P. (1988). *The Japanese: A major exploration of Modern Japan*. Truman Talley Books.

Tatar, M. (2002). *The annotated classic fairy tales*. W. W. Norton & Company.

TechSci Research. (2025). *Japan construction market—Forecast and opportunities 2019–2029*. https://www.techsciresearch.com/report/japan-construction-market/22591.html

The Association to Spread the Concept of "Sanpoyoshi" Worldwide. (2010). *What is Sanpoyoshi?* 三方よしを世界に広める会. http://sanpoyoshi.net/what_en/

The Editors. (2025). *The 5 precepts: Buddhism and morality*. Retrieved 12/01/2025, from https://buddho.org/buddhism-and-morality-the-five-precepts/

The Charter of Fundamental Rights of the European Union, (2012). https://eur-lex.europa.eu/legal-content/EN/TXT/HTML/?uri=CELEX:12012P/TXT&from=EN

Tosdal, H. R. (1936). The human factor in marketing. *National Marketing Review, 1*(3), 195–203.

Travers, J., & Milgram, S. (1969). An experimental study of the small world problem. *Sociometry, 32*(4), 425–443.

Ugander, J., Karrer, B., Backstrom, L., & Marlow, C. (2011). The anatomy of the Facebook Social Graph. *CoRR (Computing Research Repository), abs/1111.4503*. http://arxiv.org/abs/1111.4503

Urban Sustainability Directors Network. (2014). *Redefining prosperity*. Urban Sustainability Directors Network. Retrieved 25/07/2022 from https://sustainableconsumption.usdn.org/economy/redefining-prosperity

VanWolferen, K. (1989). *The enigma of Japanese power: People and politics in a stateless nation*. Macmillan.

Veblen, T. (1998/[1899]). *The theory of the leisure class: An economic study of institutions*. Prometheus Books.

Waldron, J. (Ed.). (1984). *Theories of rights*. Oxford University Press.

Wall Street Journal. (2022). *Rivals Argue Musk's Starlink satellites are overcrowding space*. Dow Jones Group. https://video-api.wsj.com/api-video/player/v3/iframe.html?guid=BB524542-5141-41E6-A3D2-0DDD72594E90

Wang, L., Schuetz, C. G., & Cai, D. (2021). Choosing response strategies in social media crisis communication: An evolutionary game theory perspective. *Information & Management, 58*(6), 103371. https://doi.org/10.1016/j.im.2020.103371

Wang, Y., Huang, Q., Davison, R. M., & Yang, F. (2021). Role stressors, job satisfaction, and employee creativity: The cross-level moderating role of social media use within teams. *Information & Management, 58*(3), 103317. https://doi.org/10.1016/j.im.2020.103317

Weber, M. (1978/[1922]). *Economy and society: An outline of interpretive sociology* (Vol. 2). University of California press.
Wenar, L. (2023). Rights. In E. N. Zalta & U. Nodelman (Eds.), *The Stanford encyclopedia of philosophy* (Spring ed.). Stanford University.
Whig. (2012). In praise of consumerism. *Adam Smith Institute*. http://www.adamsmith.org/research/articles/in-praise-of-consumerism
Wieser, H. (2016). Beyond planned obsolescence: Product lifespans and the challenges to a circular economy. *GAIA—Ecological Perspectives for Science and Society, 25*(3), 156–160. https://doi.org/10.14512/gaia.25.3.5
Wiking, M. (2017). *The Little Book of Hygge: Danish secrets to happy living*. HarperCollins.
World Health Organisation. (2022). *Constitution*. WHO. Retrieved 23/07/2022 from https://www.who.int/about/governance/constitution
World Health Organisation. (2025). *WHO COVID-19 dashboard*. World Health Organisation. Retrieved 11/01/2025 from https://data.who.int/dashboards/covid19/deaths
Worldometer. (2024). *GDP by Country*. Retrieved 21/08/2024 from https://www.worldometers.info/gdp/gdp-by-country/
Xiao, H., & Proverbs, D. (2002). Construction time performance: An evaluation of contractors from Japan, the UK and the US. *Engineering Construction and Architectural Management, 9*, 81–89. https://doi.org/10.1046/j.1365-232x.2002.00236.x
Xu, J. (2020). Does the medium matter? A meta-analysis on using social media vs. traditional media in crisis communication. *Public Relations Review, 46*(4), 101947. https://doi.org/10.1016/j.pubrev.2020.101947
Yagi, T., & Yamash'ta, S. (2022). Polyphony as a system for delivering co-creation and the empowerment of individuals. In S. Hill, T. Yagi, & S. Yamash'ta (Eds.), *The Kyoto Post-COVID Manifesto for global economics: Confronting our shattered society* (pp. 45–62). Springer Nature Singapore. https://doi.org/10.1007/978-981-16-8566-8_3
Zara, O. (2009). *Réussir sa carrière grâce au personal branding: gérer son identité et sa réputation professionnelles*. Eyrolles.
Zarkada-Fraser, A., Skitmore, R. M., & Runeson, G. (1998). Construction management students' perceptions of ethics in tendering. In J. Yang & W.-L. Chang (Eds.), *Building education and research* (pp. 80–89). E&FN Spon.
Zarkada-Fraser, A. K. (2000). A classification of factors influencing participation in collusive tendering agreements. *Journal of Business Ethics, 23*(2), 269–282.
Zarkada-Fraser, A. K., & Fraser, C. (2002). Store patronage prediction for foreign owned supermarkets. *International Journal of Retail and Distribution Management, 30*(6), 282–299.
Zarkada, A. K. (2021, June 24–26, 2021). *Listening to a unicorn's pink narrative: Scheduling of brand Posts on Instagram for maximum consumer brand engagement*. 19th International Conference on Research in Advertising (ICORIA) virtual conference.
Zarkada, A. K., & Polydorou, C. (2014). You might be reputable but are you 'Liked'? Orchestrating Corporate Reputation Co-Creation on Facebook. In T. Bondarouk & M. R. Olivas-Lujan (Eds.), *Advanced series in management—Social media in strategic management* (Vol. 11, pp. 87–113). Emerald. https://doi.org/10.1108/S1877-6361(2013)0000011009
Zhang-Zhang, Y., & Kikkawa, T. (2023). *People centric innovation ecosystem: Japanese management and practices*. Cambridge University Press. https://doi.org/10.1017/9781108981200
Zimand-Sheiner, D. (2024). Four dimensions of brand storytelling: Framework for Managing And Analyzing online brand stories. *International Journal of Business Communication, Preprint*. https://doi.org/10.1177/23294884241261362
Ελληνική Βιβλική Εταιρία. (2003). Πράξεις των Αποστόλων. In *Η Καινή Διαθήκη*. Ιερά Μητρόπολις Νέας Σμύρνης.
Ηρόδοτος (Heredotus). (1836/[450 BCE]). *Ιστορία (History)* (Α. Α. Ραδίνος (Radinos), Trans.; Vol. 1. Κλειώ). Τυπογραφία Ανδρέου Κορομηλά.
Καζαντζάκης (Kazantzakis), Ν. Ν. (1985/[1927]). *Ασκητική (Askitiki): Salvatores Dei*. Εκδόσεις Καζαντζάκη.

Τόλιος (Tolios), Γ. Υ. (2021). 30 χρόνια από την κατάρρευση του «υπαρκτού» στην ΕΣΣΔ (30 years form the collapse of "real socialism" in the USSR). *efsyn.gr*. Retrieved 05/12/2022, from https://www.efsyn.gr/stiles/apopseis/307789_30-hronia-apo-tin-katarreysi-toy-yparktoy-stin-essd

Value and Values Make Organisations Sustainable

3

The Philosophical Premise of the Discussion in This Chapter Is the Protagorean

> πάντων χρημάτων μέτρον ἐστίν ἄνθρωπος, τῶν μέν ὄντων ὡς ἐστιν, τῶν δέ οὐκ ὄντων ὡς οὐκ ἔστιν. (Πλάτων (Plato), 2015/[369 BCE])

which has been translated (Bonazzi, 2024) as "of all things the measure is man: of those that are, that they are; and of those that are not, that they are not". This widely used translation, however, masks certain important aspects of the philosopher's thought. It downplays both the judge of métron and that which is judged.

The concept of 'ἄνθρωπος' is very narrowly represented in this commonly used translation by the word 'man'. The word 'ἄνθρωπος', however, literally translates as 'human'. Contemporary philosophers (Mansfeld, 1981), as well as Plato and Aristotle (Bonazzi, 2024), concur that what Protagoras sets as the métron, the measure of all things, is not the transient and impulsive whim of a person but the understanding gained through personal experience and being a part of the past and the future of a community with a history, shared social and semantic capital (Floridi, 2018), and collective aspirations and expectations. In this sense, the métron of what is and what is not is set by humanity.

Another narrow translation from Greek to English is that of the word 'χρῆμα' as 'thing'. The word 'χρῆμα' Protagoras choses from a pool of many words that could be used to denote a 'thing' implies

(a) something that is useful, not just an object but that which can be used by someone or something that serves a specific purpose,
(b) the wealth or property one possesses, and the material resources a person or a community controls or has access to,

A. K. Zarkada and Y. Melanthiou, *Future Management for Sustainable Prosperity*, Future of Business and Finance, https://doi.org/10.1007/978-3-032-15868-0_3

(c) that which can be exchanged for something else which is useful or valued and what you receive in exchange for something else or for doing something, i.e. both the price and the cost of things or actions,
(d) money, coins, and a treasure,
(e) an exclamation used to express surprise or admiration for a large quantity or an unusually large number or volume, and
(f) an event or occurrence.

It is obvious that the very narrow translation of "*πάντων χρημάτων μέτρον ἐστίν ἄνθρωπος*" as "of all things the measure is man", what has come to be known in contemporary philosophy as the 'Man Measure' (Bonazzi, 2024), superimposes a positivist, essentially Democritean, interpretation of the Protagorean humanistic relativism. This is exactly what the philosopher and painter Salvator Rosa forces us to consider in his Baroque painting *Democrito E Protagora.* Democritus and Protagoras are debating the nature of reality, the validity of human perception, and the value of materialism. Democritus believed in a universe made of indivisible particles of matter, which he called atoms, one existing independently of humanity and regardless of the will and judgement of people. Known as the laughing philosopher, Democritus (seated alone on the right) has an amused expression and a pose of philosophical detachment reflecting his rational materialism and an atomistic worldview. Protagoras (standing over him in the centre) has a tense posture and an earnest face representing the sophists' rhetorical urgency in his challenging materialism and advocating for the people and the importance of their perception of reality.

Adopting the Democritean approach logically leads to the depersonalisation of value, a belief in the dominance of natural laws, and a subjugation of humanity to a system of atoms and void. It has dominated socioeconomic thought and practice for centuries now and has naturally led to capitalism. Under such as prism, "πάντων χρημάτων μέτρον ἐστίν ἄνθρωπος" is dismissed as the 'Man Measure'; potentially a solipsism (Bonazzi, 2024). Alternatively, it can be understood as the value métron; a statement of humanity is the arbiter of value. This human-centric perception of value is in no way equivalent to the Judeo-Christian anthropocentricity, which considers human beings as the most significant entity of the universe. It just implies that it is people who perceive something as useful or useless, so it is the human métron that sets the value of objects and acts.

Through our individual and collective experience, we set the métron of value, and whatever we do or make has value when it serves humans and the humanity which respects its past, enjoys its present, and does not sacrifice its future for a short-sighted present gain. Embracing the rights of the generations necessitates the respect and conscious and systematic protection of all human societies with which we share the planet now, and those that will come, as well as the other inhabitants of the planet and the universe which have no voice and no decision-making power in the current socioeconomic system. And it is this application of the value métron which advances sustainable prosperity for all.

In the previous chapter, we discussed métron with an emphasis on the 'mean' and 'moderation' dimensions of its meaning. We examined métron as the exercise of virtue and presented ways of finding and sustaining organisational métron to achieve legitimacy. We have demonstrated how organisations which enjoy pragmatic, cognitive, and moral legitimacy are the ones which thrive while contributing to restoring métron in the global economy and their local societies, thus advancing the cause of sustainable prosperity for all. In this chapter, we focus on sustainable organisations. We explore the concept of métron as the 'measure' of an organisation's pragmatic legitimacy, with an emphasis on the exchange dimension of pragmatic legitimacy, which is the *sine qua non* of organisational existence. We also discuss métron as the measure of morally legitimated organisational performance, which is the prerequisite of its sustainability.

This is a chapter about value, the métron of all things being exchanged, and values, the métron of virtuous behaviour: the bases of an organisation's success and ability to contribute to sustainable prosperity for all. It presents the concept of value from the perspectives of key stakeholders (investors, customers, employees, and society) and discusses the role of values in organisational design and management aiming at value cocreation. It explains the conceptual framework of the second pillar of the book's thesis, the need for values-driven value cocreation.

3.1 The Exchange Principle and the Centrality of the Concept of Value

Business is all about mutually beneficial exchanges. The métron of a 'beneficial' exchange is value. Value is one of the most difficult concepts to define, not only within management and marketing discourse (Edvardsson et al., 2011; Kelly et al., 2017; Mingione & Leoni, 2020; Panigyrakis & Zarkada, 2014) but also in philosophy (Scruton, 2012). Most of the time, the term is used as if it were self-evident, and it remains undefined, even when it is the central concept in research or the semantic foundation of the author's perspective (Chapleo, 2010; Grönroos, 2011; Grönroos & Voima, 2013; Healy & McDonagh, 2013; Ingenbleek, 2007; Payne et al., 2008). In the marketing literature, value denotes the experience gained by satisfying a vague desire for acquisition, or the price, benefits, and relative significance of something. Often, it is equated with abstractions such as quality or image. Thus, value is seen as nothing more than the automatic outcome of the production and commercialisation process; the better they function, the higher the value they produce for the parties to the exchange process.

Value is generally accepted as being at the core of an organisation's competitiveness or attractiveness to investors, and in this sense it is offered as a proxy of the efficacy of its pricing policy and of its total worth. Yet, management scholars, just like their marketing colleagues, struggle to define value or propose methods for measuring it (Fotiadis et al., 2022; Mahajan, 2017a; Weinstein, 2018; Weinstein & Johnson, 2020). Nevertheless, "[o]ne of the crucial advances in any research field is to clearly define its central construct" (Agle & Caldwell, 1999), and to this we aim to contribute with this discussion. We adopt the perspective that "[t]he value concept.. represents a link between society, markets and companies and dominates decision-making, the companies' success and the wellbeing of people involved in business contexts" (Mahajan et al., 2023). So, we echo other authors' sentiments expressed in statements such as that "[v]alue is.. improving the well-being or worth for ourselves and others" (Mahajan, 2017b) through "caring,.. progress and creativity" (Mahajan et al., 2023). Following this logic, we argue that value is the central concept of all economic (or is it all human?) activity.

Value can be individually or collectively perceived, experienced, or cocreated (Zeithaml et al., 2020). This is what Marx defined as the use-value aspect of a commodity: its ability to provide satisfaction through the alignment of its properties with the perceiver's, user's, or owner's needs and wants (Varul, 2011). Use value can be inherent to an object, such as the air we breathe, or derived from the expenditure of materials, skills, and effort (Marx, 2018/[1867]). This transformation of raw materials and ideas into useful objects is of a qualitative nature and as such is not a measure of anything but someone's abstract satisfaction (Varul, 2011). The same objects, however, acquire a measurable quantity of value when exchanged. This is the exchange value of a commodity which is "independent and definite" as it is equated with the independent and definite value of another object (Marx, 2018/[1867]; Varul, 2011).

Value materialises through exchange. When a value-bearing object is exchanged, the use value the owner enjoyed is transformed from a subjective evaluation and a psychological state (Schroeder, 2021) to an objective measure of the object's worth to the receiver in the form of the object the receiver gives in exchange. The use value of the house I inherited from my grandfather comprises the importance I place on my childhood memories, my enjoyment of it as a place to live in, and the aesthetic pleasure its architectural features provide me with. Its use value is perceptual and emotional, and it is all mine and not measurable. Its exchange value is not what I think I will get when I put it on the market, as many people believe, but what someone pays me to move out and let them move into my grandfather's house. The exchange value is objectively measured: it is a luxury sports car or a smaller house in a more fashionable neighbourhood.

For an exchange to occur, there must be at least two voluntarily participating parties which are free to accept or reject the other's offer. Each party must possess something of value to the other party (in the context of market capitalism in its basic contemporary form, the one party is a seller who is in possession of an object and willing to part with it and the other party is a buyer who is in possession of money and willing to spend it). The two parties need to be able to communicate and make their respective offers available. The exchange will take place when both parties believe that it is appropriate or desirable to engage with the other party and they perceive what is being offered as of sufficiently high value to justify the sacrifice of the value they already possess, as well as the effort and risks involved in the transaction (van Waterschoot & Van den Bulte, 1992). In this sense, value is what the sellers—and here we use the term 'seller' loosely to denote any entity seeking to enter into an exchange relationship with another party, such as a manufacturer seeking to sell its products, a celebrity seeking the audience's attention that can be monetised through endorsements, or a charity seeking donations from the privileged classes to finance its services to disadvantaged communities—get in exchange for what they offer from the other party—loosely represented by the term 'buyer'.

When the exchange conditions are in place, the exchange does not take place automatically. They are necessary conditions but not necessarily sufficient ones. A number of barriers need to be overcome by management actions using marketing instruments (van Waterschoot & Van den Bulte, 1992). These can be information barriers (e.g., lack of awareness or insufficient information to support the seller's pragmatic legitimacy), credibility and perceived risk barriers (i.e., limited or no moral legitimacy of the seller which results in lack of trust and exaggeration of the risks inherent to the transaction), competitive barriers (such as the availability and attractiveness of substitutes (van Waterschoot & Van den Bulte, 1992)), and physical and psychological barriers experienced by the sellers which can be anything from inertia and habits to tastes and preference structures (van Waterschoot & Van den Bulte, 1992).

If we look at exchange from the perspective of the sellers, the exchange principle dictates that they perform four 'generic functions', usually assigned to marketing (van Waterschoot & Van den Bulte, 1992), but actually permeating the organisation (Webster Jr, 1992):

- formation (of an object, which can be a physical object, an idea, a service, an event, or anything that can be exchanged, used, enjoyed, or consumed, and which is perceived as valuable by the other party, which we schematically call the seller),
- valuation (assigning the object an estimation of the value the seller will receive in exchange for it vis-à-vis the total sacrifice the buyer will make to acquire it),
- facilitation (designing and overseeing the transaction process by making the object available to the seller), and
- symbolism (drawing the seller's attention to the objects' features and meanings and creating cognitive, affective, and conative bonds which affect preferences and influence behaviours).

These functions are not natural phenomena which take place automatically. They need people to perform them, physical objects to carry them out (such as offices, retail spaces, tracks, and billboards) and to convey the symbolisms of them (such as staff uniforms, colours, textures, smells, and sounds), and systematised processes that ensure all the functions are performed efficiently and effectively. So, to complete the list above we here add:

- integration (orchestrating all the functions to produce a predetermined result in the form of a coherent and distinct value-bearing entity such as an object, a service, an idea, or an experience).

These functions correspond to the various elaborations of the classical Marketing Mix (with its 4Ps (McCarthy, 1960) or 7Ps (Wirtz & Lovelock, 2016) variants and its improved classification of instruments (van Waterschoot & Van den Bulte, 1992) and the 7Es of its experiential version (Batat, 2022). In essence, they are different approaches to the exchange value management process (Table 3.1).

Table 3.1 Synthesis of the various approaches to exchange value management process

Function	Marketing Mix variant	Elements	Instruments (van Waterschoot & Van den Bulte, 1992)		Experiential marketing mix Dimensions (Batat, 2022)
Formation	4Ps (McCarthy, 1960)/7Ps (Wirtz & Lovelock, 2016)	Product	Basic Mix		Experience
			Promotion Mix		
Valuation	4Ps (McCarthy, 1960)/7Ps (Wirtz & Lovelock, 2016)	Price	Basic Mix		Exchange
			Promotion Mix		
Facilitation	4Ps (McCarthy, 1960)/7Ps (Wirtz & Lovelock, 2016)	Place	Basic Distribution Mix		Extension
			Promotion Distribution Mix		
Symbolism	4Ps (McCarthy, 1960)/7Ps (Wirtz & Lovelock, 2016)	Promotion	Basic Communication Mix	Mass Communication Mix	Emphasis
				Personal Communication Mix	
				Publicity Mix	
			Promotion Mix	Mass Communication Mix	
				Personal Communication Mix	
				Publicity Mix	
Integration	7Ps (Wirtz & Lovelock, 2016)	People			Empathy capital
		Physical Evidence			Emotional touchpoints
		Process			Emic/etic process

We here argue that all economic activity is nothing but people performing their allocated integrative roles with respect to the formation, valuation, facilitation, and symbolism of value. Subsequently, social structures reflect the division of labour and the control over the resources required for value exchange. Thus, value is at the core of the socioeconomic system and the métron of its efficiency with respect to division of labour, control of resources, and distribution of responsibilities, gains, and externalities. For this reason, but also because it is a critical success factor for any research to define its central concept (Agle & Caldwell, 1999), we come back to the meaning of value. We look at both real definitions of value (Locke, 1690), that is, statements of its an object's essential attributes that form its essential nature (Αριστοτέλης (Aristotle), *trans.* 2015/[fourth century BCE]) and nominal ones (Locke, 1690), which explain its meaning, like a lexicographer does (Gupta, 2021). It is obvious that both aspects of the definition of value are dependent on the perspective of the observer. As John Locke (Locke, 1690) put it, "complex ideas ... [are] combinations of simple ideas put together, and united under one general name" by our mind, which "uses some kind of liberty in forming those complex ideas". Thus, he explains why "one man's idea of gold, or justice, is different from another's ... because he has put in, or left out of his, some simple idea which the other has not (Locke, 1690)".

In the context of the socioeconomic system, it is not a man (or a woman, or any human sensory systems and brain for that matter) that is the observer of an object and perceiver of its value. It is rather a stakeholder group that assigns value to exchanged objects and their role in forming relationships, satisfying needs, satiating desires, and meeting personal and organisational objectives. Value is the métron of different things to different stakeholders (e.g. investors, customers, or employees of a company, or local communities, society, or natural ecosystems). So, in the next sections we attempt to answer the fundamental question: Value for whom?

3.1.1 Value for Investors

In a book called *Value-Based Marketing* (Doyle, 2009), the term 'value' appears 1,857 times without ever being explicitly defined. However, a close examination of its usage reveals the philosophical orientation of the author, which aligns fully with the dominant view of the role of marketing in the contemporary socioeconomic system and the commonly held beliefs about the purpose of economic activity: "to contribute to the maximization of shareholder value". Indeed, the métron of the success of management strategies and actions is positioned as "the value they create for investors" (Doyle, 2009). Viewed from this perspective, value is simply the dividends and profits generated for a small group of individuals who have invested their capital in the creation and operation of an organisation. Their satisfaction with the exchange comes not from feeling that they have received something commensurate with what they have given but from multiplying their offering.

The pursuit of perpetual wealth accumulation for investors is the managers' responsibility if not *raison d'être* as illustrated in the following statement: "today's managers know that unless they do this, their jobs will become vulnerable [so] earning returns that shareholders regard as acceptable is a necessity for survival" (Doyle, 2009). Even when managers strive to meet long-term organisational objectives such as the acquisition of intangible assets (e.g. strong brands, a positive corporate reputation, competitive advantage, technical know-how, etc.), even ones which are detached from short-term cost cutting and reframed as the "increase in the net present value of long-term cash flows" (Doyle, 2009), the core assumption remains unchanged: the role of management is to increase shareholder wealth.

This narrow, but pervasive in the capitalist system, conceptualisation of value inevitably contributes to the pressing global challenges of our time (such as social inequality, poverty, food insecurity, environmental degradation, and climate change) because it dictates that shareholders reap the surplus value of others' transformative labour (Marx, 2018/[1867]; Varul, 2011) while societies and the natural environment bear the full cost of the negative externalities of economic activity. Moreover, when the managers who are responsible for the formation, valuation, facilitation, and symbolism of objects which are perceived as bearing value are guided by the shareholder value maximisation goal, it is only logical that they will target wealthier social strata with high-margin offerings and withhold services that are not profitable from BoP market segments.

The focus of managerial decision-making on generating value for investors is particularly problematic when the value-bearing objects are essential for their users' survival or maintenance of a dignified and secure living. In the health sector, for example, managers pursuing shareholder value as the ultimate goal of their efforts are more likely to invest in the research and infrastructure development required for low-risk and high-return procedures, such as elective cosmetic surgery, thus depriving the general population of necessary healthcare services. Moreover, the symbolisms that support such service offerings need to promote idealised standards of beauty which affect the self-image of both those who can afford to buy the relevant services and of those who cannot. This way, the healthcare and beauty industry managers pursuing investor value generate negative externalities in the form of emotional suffering to the most vulnerable members of society who perceive themselves as doomed by their limited means to always be falling below what they are made to believe to be a societal standard of beauty. These are only some of the ways in which human rights, equality, justice, democracy, the quality of life of the poor, the urban environment, and the ecosystems end up becoming collateral damage, externalities which are excluded from the calculations of value for investors.

In the 3BL approach to sustainability conceptualisation, it is this approach to value that is reflected by the 'profits' circle (Fig. 1.7). If some of it is used for environmental protection projects, then life on earth is assumed to be made more viable. If some of it is returned to society, then the economy is supposed to be made more equitable. The fact remains, however, that the largest part of the profits pie is allocated to the few that own and control the factors of production. The net worth of the world's, or each country's, richest people, however, and the way they conspicuously or discreetly use their wealth, amply demonstrate that even if they channel some of it to charitable and environmental causes, the fact remains that it is their money and our problems.

3.1.2 Value for Employees

Take for example a recent statement by "a spokeswoman for Mr. Gates" that "Bill Gates remains as committed as ever to advancing the clean energy innovations needed to address climate change" in a press conference announcing that one of the world's richest people, who "has established himself as one of the leading voices pushing businesses and governments to address the threats of a rapidly warming world", is slashing the budget of his environmental protection charity and firing its staff (Gelles & Schleifer, 2025). This small news item about Bill Gates and his climate action highlights the disposable nature of managers in the current socioeconomic structure, irrespective of their being employed by charitable, profit-making, or profiteering organisations. Although the organisational culture ideological superstructure and corporate events craft and support a pervasive illusion that the management classes are part of the ruling elites and decision-makers, the reality is that they are not part of the value-for-investors orientation of economic activity, even when they are shareholders of the organisations that employ them.

In the management literature, value for employees is seen as a construct that links human resource management actions such as training (Glaveli & Karassavidou, 2011) and development, or at least the employee's perception that the organisation cares about their professional development (Lee & Bruvold, 2003), with job (Lee & Bruvold, 2003) or employee satisfaction (Glaveli & Karassavidou, 2011) and the higher levels of employee loyalty (Glaveli & Karassavidou, 2011; Pokorny, 2013), commitment (Lee & Bruvold, 2003), or engagement (Pokorny, 2013), which are associated with reduced employee turnover or intention to leave one's job for another (Lee & Bruvold, 2003). The value a company promises and delivers to the employees has also been seen as a way to increase the likelihood that employees will deliver the firm's promises to customers (Boukis & Kabadayi, 2020), which in turn increases customer satisfaction (Glaveli & Karassavidou, 2011), loyalty (Glaveli & Karassavidou, 2011) and, at the end of the day, profitability (Glaveli & Karassavidou, 2011). In this sense, value for employees is just a stepping stone to increasing value for investors.

Other approaches view the value for employees as derived from tangible and intangible resource exchanges between firms and employees or by the "employee's experiential perception of the value that emerges from possessing or utilizing firm resources over time" (Boukis & Kabadayi, 2020). These might be reciprocal and can take various forms, such as salaries for labour, support for team spirit, developmental opportunities for engagement, or pride for a strong corporate reputation. Thus, the service logic is extended to see the employer-employee relationship as a process of utilisation of the resources the parties possess to generate value which they then use to "acquire new resources through exchanging their personal resources" (Boukis & Kabadayi, 2020). Although in this approach the possibility of the employees ending up with negative utility is mentioned in the passing, this conceptualisation masks the immense power imbalances in such structures. Moreover, in such discourses, paid vacation time is conceptualised as use of corporate resources and training as their possession.

Regardless of the philosophical and logical objections we might have, service logic approaches, and the related internal marketing ones, are closely linked to the idea that employees should be treated as "an intermediate set of customers (i.e. internal customers), whose needs should be met first so that they better align with strategic marketing objectives and eventually satisfy external customers' needs (Boukis, 2019)". The tool of engaging employees into exchanges with emotional undertones is internal marketing, which is tasked with "value-identifying, value-generating and value-enhancing activities" which enhance employee perception of deriving value from their employment, which, in turn, results in "reciprocal responses and obligations towards the firm (Boukis, 2019)". The view that treating employees as human beings with material and emotional needs is just a means to the ultimate end of profit-making is unquestioningly accepted as a fact in conceptual works on motivation (Pokorny, 2013) and empirical studies from the perspective of psychology (Che et al., 2022) alike.

This instrumental view of people is summarised in the term 'human resources', which places human beings, with their skills, talents, knowledge, time, and effort, on the same level of utility with land and capital. Similar in its instrumentalism, but morally even worse, is the term 'human capital', which implies that the effort, knowledge, skill, and commitment of people can be accumulated, controlled, gambled, and squandered. Nevertheless, there seems to be a consensus on the link between how an organisation treats its human and social capital and its ability to yield customer value (Morris & Snell, 2019).

3.1.3 Customer Value and Value for Customers

Customer value is the measure of how much money an organisation earns from each customer or market segment. Clearly, the underlying logic of the customer value concept is a conscious divergence from the application of the exchange principle. The reason for cultivating human and social capital is not to simply give consumers at least equal value in exchange for the value they part with—their hard-earned money, time, effort, attention, etc.—but to organise relationships with them in a way that maximises the value the organisation gets, after allowing enough of the value they garner to be distributed to employees in a way that maximises their effort towards generating more shareholder value. Moreover, in this framework, the value an organisation offers its customers is the basis of its competitive advantage; that is just a tool to attract their value from competitors. Organisations focused on maximising value for investors were thus expected to be profitable in the long term at the expense of both customers and competitors (Roig et al., 2006).

This focus on shareholder value, however, was eventually recognised as detrimental to businesses as it resulted in customer dissatisfaction, which diverted customer value to competitors who were perceived by consumers to offer more for the value they had to part with to acquire their products or enjoy their services. Thus, in the marketing literature, value for customers, expressed by the concept of customer perceived value, was identified as a more solid basis for a firm's competitive advantage some three or four decades ago (Zeithaml et al., 2020), as it was linked with customer satisfaction and loyalty (Roig et al., 2009). Despite its many variants and diverse measures (Roig et al., 2006), customer perceived value has found its way to the general management and business administration texts. It is a multidimensional and dynamic concept which has been connected to consumer behaviour (in simple terms, it is the reason why someone will buy, buy again, and choose one provider over another (Mahajan, 2017a)) and, through it, to organisational outcomes, even though "it is not yet fully understood exactly what consumers want (Roig et al., 2006)".

Customer perceived value is defined as the sum of the perceived value of the acquired product or service, combined with that of the point of sale and the service staff. To measure customer perceived value, researchers propose that mental and cognitive evaluations of economic and functional utility are taken into account (e.g. how clean the store was, whether the price is justified by the quality, and whether the product delivered what the seller promised), as well as emotional reactions (e.g. how pleasant the shopping experience was or how friendly the shop owner was), and social judgements (e.g. what my friends will think of me when they see me using the product) (Roig et al., 2006; Verhoef et al., 2009). But, in exchange relationships, one form of value is given up to acquire another. This means that the standardised mental process considers the benefits the exchange provides (solving practical problems, creating a sense of security, improving the consumer's social image, etc.) after subtracting the cost of acquiring and using the product (money, time, effort, etc.) (Kelly et al., 2017; Mahajan, 2017a; Parasuraman et al., 1985; Weinstein, 2018; Weinstein & Johnson, 2020; Zeithaml et al., 1993). This conceptualisation is generally accepted, even by scholars adopting a socialist perspective and examining Bottom of the Pyramid markets (Srivastava, 2019).

In this framework, marketing provides information, cues, and symbolism to make customer-perceived value appear to be as large as possible. The more convincingly the product benefits are presented, and the more effectively the consumer-perceived costs and risks are addressed, the greater the perceived value will be (Daryanto et al., 2010). The role of management is to remove or, at least, manage actual barriers to consumer action (such as information, competitive, or physical barriers), or those related to consumer perceptions (such as credibility, perceived risk, and psychological barriers) (van Waterschoot & Van den Bulte, 1992). For this, they need to mobilise and harness firm capabilities, such as market sensing (identifying and evaluating prospect and customer perceptions of value), seizing opportunities (committing resources to deploy value to customers), and transforming the structures of capturing value (Teece, 2007). So, in summary, the expected result of effectively communicating a positive benefits-barriers balance is sustainable superior firm performance (Chauhan, 2024; Paul & Rosenbaum, 2020; Teece, 2007), which results from continuously "delivering value to customers and capturing value" (Paul & Rosenbaum, 2020).

3.2 Ethical and Conceptual Problems of Focusing on Value for a Specific Stakeholder Group

Viewing investor profit through internal and external customer value exchanges as the sole purpose of business activity is morally problematic not only in terms of judging the intention, but also when examining the means, the actual actions it drives managers to take, and when assessing the, one could argue, disastrous consequences for the environment and social justice. Even if one adopts a totally amoral standpoint (i.e. without making ethical or evaluative judgements), defining value as synonymous with investor, employee, and customer-perceived value creates practical problems. The fact that value remains an implicit and indirectly defined concept makes it hard to understand, systematically study, and measure. Moreover, viewing value as "idiosyncratic, experiential, condition-dependent, and meaning-laden" (Edvardsson et al., 2011), or as a transient, general, and undefined preference related to an interactive experience (Zeithaml et al., 2020), renders it entirely subjective (Parasuraman et al., 1985). So, value becomes a purely personal assessment by the investor, the employee, or the consumer of the degree to which they have achieved their egoistic goals (Holbrook, 1987). This means that the other party to the exchange process, for example, the seller seeking to deliver customer value in the most basic buyer–seller relationship, cannot have a clear enough picture of the goal they are to meet (Roig et al., 2006).

The academic fields that aim to address this ambiguity and overcome the barrier of total subjectivity are those of organisational and consumer behaviour. Both try to make exchange-related phenomena—such as engagement with one's work or commitment to an employer and attachment to brands or consumption patterns—understandable as psychological and social processes. They have both made undeniable contributions towards understanding the social and cultural influences on human behaviour and have provided insights into the psychological, experiential, symbolic, and ideological dimensions of work and consumption. In the past century or so, however, they have increasingly become so broad and their findings so diverse and, often, conflicting that they obscure more than what they clarify (Arnould & Thompson, 2005; Holbrook, 1987). This is partly the reason why they have, so far, failed to provide a foundation for effective policymaking to limit overconsumption and control environmentally harmful production processes (Wilk, 1999, 2002).

The other reason why organisational and consumer behaviour research has failed to control the socially and environmentally harmful consequences of the exchange processes, however, is that their objective is not knowledge for the sake of knowledge, or to be used for improving societal processes and human well-being. The objective is to find ways to manage processes and their participants in a way that increases investor profits. On rare occasions research on organisational and consumer behaviour is concerned with improving employee well-being or preventing harmful consumption and promoting socially beneficial behaviours. Even in these cases, however, the underlying motivation is linked to profits for shareholders or cost reductions in the form of, for example, reducing absenteeism or public health costs, which, again, end up directly or indirectly benefiting capitalists, for example, by lowering taxation rates.

Capitalists' most powerful tool is classical marketing. Marketing not only propagates consumerism but has managed to elevate it to dominant ideology status (Maycroft, 2004; Woodall, 2012), which supports the edifice of modern globalised capitalism (Schmitt et al., 2021) and "the commercialization of everyday life" (Granot et al., 2014). Capitalism views markets as inherently fair and meritocratic. Market fairness is premised on the disconnect between the producer (e.g. the wheat farmer), the product (e.g. prepacked sliced bread), and the consumer (e.g. the household) within the system of compartmentalised intermediaries to the exchange mechanism (e.g. flour mills, wholesalers, food industries, and supermarkets) (Levine, 2007). Market meritocracy is premised on the prevalence of the most capable competitor within a system of unrestrained competition. The fairness and meritocracy of the markets are expected to result in better products and prices for all participants in the capitalist system. These are deeply held beliefs widely shared not only among the ruling financial elites but also the intelligentsia, the precariat of young professionals, and even the working classes. They are so pervasive that they mask both the functioning of markets as mechanisms of oppression (for example, by reinforcing ethnic and racial divisions through brand messaging (Crockett, 2021) or through the commodification of sexual violence by the pornography industry (McVey et al., 2021)) and the potential of marketing to serve as an agent for positive change.

Even consumer- and human-centric marketing are presented as tools for building attraction for highly differentiated and, as such, more profitable brands (Kotler et al., 2020). In this sense, they are also powerful instruments of consumer capitalism. Consumer capitalism is not concerned with community solidarity, personal spirituality, or the ethical integrity of the system of wealth production and distribution (Levine, 2007). Thus, environmental pollution from factories, marine ecosystem disruption due to overfishing, the exploitation of the vulnerable, poverty, and other consequences of consumerism are understandably outside the scope of a discipline focused on easing the "pains and difficulties imposed by prices and budget constraints ... and by an existential anguish in choosing among products, none of which is perfect" which "modern consumers face" (Holbrook, 1987).

In the early days of contemporary market capitalism—in a world devastated by World War I, suffering the consequences of the 1929 Crash, and on the brink of World War II—the "challenge to modern business [was] to discover how to get Mr. and Mrs. Consumer to consume ... to encourage the maximum consumption of goods" so as to enhance "the profit-making possibilities for business" (Hess,

1935). Half a century later, marketing was expected to restore 'Mr. and Mrs. Consumer' of the 1930s to the "State of Paradise in which Adam and Eve's sole task was to enjoy pleasant forms of consumption" (Holbrook, 1987). Almost another fifty years later, most households do not enjoy blissful consumption in their personalised and Instagrammable Gardens of Eden. Instead, the inhabitants of this small planet in a small galaxy are faced with the problem of rising mounts of landfill rubbish and will soon have to do something about the increasing amounts of dangerous space debris which our successfully maximised consumption has been producing.

Never has the Protagorean value métron been more in need of application at the level of 'Mr. and Mrs. Consumer' (Hess, 1935), of 'Adams and Eves' (Holbrook, 1987) who live only to consume as much as possible in order to gain social status and express their individuality (Levy, 1959). We need a value métron which embraces all of humanity and our habitat. We need to focus on value for all. And we need to assume responsibility for placing the pursuit of value for all in the place of a guiding principle for Future Management. This requires a mindset regeneration. Such profound change cannot take place as a result of successful communications campaigns. What we know from the in-depth study of the destructive phenomenon of panic buying (Billore & Anisimova, 2021), persuasion does yield quick short-term results, but it takes a sustained and systematic two-way communication effort to bring about the integration of all stakeholders into a vision of effective and efficient social change.

3.3 Value for All Is a Values-Driven Choice

The twenty-first-century literature has started to distinguish between everyday value (delivered by organisations through the market exchange mechanism to meet context-specific needs) and transformative value, which has a clear social dimension and is cocreated when market actors are consciously and actively engaged in transforming the historical, cultural, and social contexts in which they (Parsons et al., 2021; Zeithaml et al., 2020). Cocreating transformative value requires that market participants are fully aware of their role in the reproduction of institutionalised regimes and the evolution of organic regimes (i.e. those emerging through customs and behavioural norms). To cocreate transformative value, the participants in the exchange process need to challenge the status quo, seek an escape from the regimes they question, and act based on the values they embrace. Ultimately, the criterion for recognising a value as transformative is the improvement in the quality of life and the holistic increase of well-being for individuals and social groups without disadvantaging other individuals and groups.

Recent streams of thought such as Transformative Services Research (TSR) (Parsons et al., 2021) and the study of ideologically driven consumer behaviour (Schmitt et al., 2021) are gaining ground, even if their voices are still weak compared to the dominant capitalist worldview. Their key takeaways are that businesses produce transformative value when they adopt high environmental production standards, take responsibility for collecting and recycling their products and packaging materials after consumers dispose of them, pay fair wages, conform to safety regulations, engage in social or philanthropic initiatives, etc. Examples of transformative value-generating consumer behaviours include ethical consumption, boycotting values-violating businesses, purchasing fair trade products, active participation in the circular economy etc. In both cases, transformative value is cocreated by practicing citizenship and employing métron. It is obvious that transformative value is inherently universal as it is impossible to produce at the expense of people, social groups, or the environment. Value for all is also inherently sustainable and its métron is how close it brings the world to sustainable development for all. It is also inherently values-driven.

Values (what is truly good, desirable, and so important as to guide the course of our lives, such as justice, altruism, equality, peace, and a beautiful and safe world (Pelsmacker et al., 2005)) are collective, mostly universal, and timeless. Their force and prioritisation may be subjective—as many have suggested Protagoras implied (Bonazzi, 2024; Mansfeld, 1981)—but their truth and validity are objective—as Plato (Plato, 2008) emphasised. Values form the foundations of social organisation as they represent the métron of socially and morally acceptable behaviour. For organisations, their values statements represent their collective truths about what is important and what people in the organisation must do even if they sometimes do not want to (Scruton, 2012). In the organisational and consumer behaviour research, as well as in the social psychology literature, such values are sometimes referred to as terminal values in order to distinguish them from instrumental values, which are merely statements of means for achieving goals (Agle & Caldwell, 1999; Pelsmacker et al., 2005; Shaw et al., 2005). This distinction, however, is both unclear and impractical. It is more practical—and also a common practice in philosophy (Scruton, 2012)—to use the term values exclusively for terminal values, while instrumental values, due to their subjective dimension, are better referred to as value in the sense of Protagoras' 'χρῆμα', a useful thing, i.e. something that has utility for someone, under certain conditions, for a specific purpose.

3.3.1 From Needs and Desires to Value

To humanise capitalist businesses, the management and marketing literature puts forward the satisfaction of needs and desires through consumption as their overarching purpose. Consumer research tries to legitimise this notion by striving to understand what people consume and why. A central theme in this extensive (Wilk, 2002) yet excessively complex and inconclusive (Tadajewski & Jones, 2011) body of knowledge is the debate around the distinction between 'real' needs that must be satiated and 'false' desires, the satisfaction of which ought to make life pleasurable. This discussion largely reflects pervasive philosophical perspectives, such as the Aristotelian problematisation regarding appetite, will, and desire, or the Cartesian distinction between body and mind (Panigyrakis & Zarkada, 2014). Generally, the 'reality' of needs is positioned in terms of their reflecting 'natural tendencies' which can be satiated (e.g. hunger that ceases once we have eaten enough) (Galbraith, 1998/[1958]; Scanlan, 1967; Smith, 1761), while desires are labelled as 'false' because they are seen as emerging from various external or psychological factors (such as the devil in Islamic teachings (*The Holy Qur'an: Translation*, 2000), society (Scanlan, 1967) and capitalism in Marxist thought (Petrovic, 1991), or our imagination (Scanlan, 1967)). The characteristic of desires is that they are considered to be not only insatiable but also increasing; the more you have, the more you want, like wealth and power, which are never enough for those pursuing them (Galbraith, 1998/[1958]).

While philosophers and economists hotly debate this distinction, businesses use marketing to bypass it. This is evident in the literature where terms like "needs and/or wants," "needs, wants, and desires," or "needs, wants, desires, and expectations" have been used since its beginnings (Benson, 1937; Churchill, 1935; Crossley, 1937; Nash, 1937) and until the turn of the twenty-first century. Recent approaches to understanding consumption, like Consumer Culture Theory (CCT), seem to return to the Aristotelian-Cartesian foundations of classical economic theory in their seeking to identify consumption patterns and motives in terms of gender and sex differences, race, religion, or nationality, age group and generational cohort, football team identification, and political party membership, thus creating new arbitrarily dichotomous systems akin to the need/desire and body/mind (or soul) distinctions. Similarly, consumption decision-making is viewed as a problem of bounded rationality, like a set of choices between price and functionality within given cultural, social, and psychological dynamics (Arnould & Thompson, 2005; Hungara & Nobre, 2021; Simon, 1955, 1979).

Yet, the postmodern experience people in the advanced economies share is our being increasingly multidimensional entities—both as individuals and as collectivities. What better describes contemporary people and peoples, each of us and all of us together navigate through a continuous sequence of identity choices, identities drawn from a multitude of cultures, bodies of knowledge (Rotaru et al., 2010), and traditions, and used or discarded depending on context, if not mood of the moment. Each of the identities we come to choose from, or combine and morph into, carries meanings and its own validity, none dominates (Anderson, 1995). These identity projects challenge various orthodoxies (van Raaij, 1993), embrace and celebrate plurality, and supersede Manichaean dilemmas. Thus, all needs, wants, and desires, regardless of whether they are real or imaginary, innate or socially and class-constructed (Gasana, 2009; Veblen, 1998/[1899]), unspoken and bodily felt or expressed through narratives and imagery (Bourdieu, 1984; Rotaru et al., 2010), can be accepted as equally valid. The objects of consumption related to them are instrumental to the definition, maintenance, and communication of identities, and in this sense they comfortably fall under concept of value (Panigyrakis & Zarkada, 2014).

3.3.2 Value Cocreation

The problem of value production, that is, which is the nature of the value being produced, where, when, how, and by whom is value produced, acquired, consumed, or destroyed, is one with which classical and neoclassical economic theories have grappled for decades. Yet, it remains insoluble. The underlying assumptions of economic theories reflect the realities of industrialisation and modernity: value is (a) determined by the producer and incorporated into the product through the production process, or (b) defined through the exchange process, or (c) created by the consumer during use, and (d) destroyed by use or consumption. These conditions, however, hardly reflect the realities of economies of knowledge, information, and services, let alone the experience of the constantly connected, multitasking, and media meshing people of the 2020s who lead a phygital existence across time and space boundaries, if not realities. The sharing and platform economies are cases in point.

One solution to the conundrum of value production and consumption has been the repositioning of value as being coproduced (Kokkinopoulou et al., 2024; Muniz & Guzmán, 2023; Vargo & Lusch, 2004; Yi, 2014) when all actors play their predetermined roles and, sometimes, codestructed, when one or more actors fail to act in their predetermined boundaries and prescribed manner (Bu et al., 2022; Iglesias et al., 2022; Kashif & Zarkada, 2015; Lumivalo et al., 2024; Plé & Cáceres, 2010). In the context of the phygital economy, value is not based on production, transaction, consumption, or necessarily on a physical object. Value is coproduced by a multitude of material and intangible factors, a network of actors—who can be producers, or producers and retailers, customers and users, or who can play multiple roles—and resources (Löbler, 2013; Vargo & Lusch, 2011). Value is cocreated by the interaction of actors, operand resources (the tangible, static, and typically passive classical economic resources such as land, raw materials, machines, etc., which are acted upon to produce an effect), and operant resources (the intangible, dynamic, and active resources such as knowledge, routines, relationships, etc., which act upon other resources and transform them into value-bearing entities) (Vargo & Lusch, 2004).

Managers design the systems and mobilise the necessary operant and operand resources and actors (who can also act as resources (Löbler, 2013; Vargo & Lusch, 2011)) for value cocreation and perform the service script which integrates them. Value is thus cocreated, but when resources are misused by any system participant, value is codestructed (Kashif & Zarkada, 2015; Plé & Cáceres, 2010). Nevertheless, value is jointly and mutually cocreated at every stage of the design, production, exchange, and experience process through the interactions of network actors and resources and requires the commitment of skills and competences of all involved (including those traditionally described as customers or end-users who also contribute not just money but also skill, ideas, commitment, time, etc.). Thus, value is always the result of a dialogue about what the mutually beneficial outcomes are.

Taking this understanding a step further, it is clear that value bears "polysemous meanings" (Mahajan et al., 2023) comprising functional, as well as cognitive, emotional, and self-expressive dimensions (Mingione & Leoni, 2020). This view, known as Service-Dominant (S-D) Logic (Löbler, 2013; Lusch & Vargo, 2006; Vargo, 2009; Vargo & Lusch, 2004, 2018), solves the problems of conceptualising value in the context of the twenty-first century phygital markets and within the current form of capitalism but does not solve the problem of negative externalities.

3.4 Value Without Negative Externalities

The range and number of objects of exchange and consumption (such as branded products, commodities, gadgets, services, celebrities, identities, narratives, tourism destinations, commercialised experiences, etc.) have increased exponentially in recent years, but so have the range and number of market externalities (such as air pollution, mounts of rubbish, and deforestation coupled with corruption, excessive wealth concentration, and social exclusion, to name but a few of the hardest to control at the moment). Even people's private lives are objects of externality-producing exchanges. In the social media markets, for example, people's friendships, sorrows, opinions, food, and clothes become monetised data; they are exchanged for the fleeting reactions of next-door neighbours or far-away strangers, and generate externalities such as cyberbullying and addiction (Dwivedi et al., 2021), digital fraud, disinformation, or excessive energy and water use (Shein, 2025). Finally, we should not forget the negative externalities of exchange relationships which are experienced by individuals, like affluenza and its physical, mental, and social health symptoms which we discussed in [[Ch1_Permacrisis]].

The common thread linking all negative externalities of market exchanges is the lack of métron. Overconsumption degrades the physical environment as well as human health, while practicing métron in the form of temperance (as we discussed in [[Ch2_Sustainable prosperity]]) reduces the number and extend of negative externalities. This, of course, is not just common sense; it is the cornerstone of various cultural and religious traditions as well as political ideologies and forms of socioeconomic organisation. It is, however, contrary to the fundamental logic of contemporary capitalist enterprises, the neoclassical view of economic theory, and the economic development school of thought who systematically promote a narrow interpretation of utilitarianism and a maximalist reading of individualism (MacIntyre, 2007; Russell, 1946/2004) which link prosperity with the acquisition of material goods through markets which ensure the maximisation of individual benefit, even at the expense of the collective good. In the past couple of decades, the number of critics of such a worldview has been increasing (Coffin & Egan-Wyer, 2022; Shankar & Fitchett, 2002; Shankar et al., 2006) and the empirical evidence demonstrates that capitalism, and its key tool, marketing, cannot keep the promise of happiness through having, consuming, or showing off possessions (Coffin & Egan-Wyer, 2022; Shankar et al., 2006). Thus, twenty-first-century business and marketing scholars agree with psychologists (Csikszentmihalyi, 1999;

Van Boven & Gilovich, 2003) and philosophers (Fromm, 2013; Marcel, 1965) who link a life focused on *having* material possessions and power with the greed, envy, and violence (Fromm, 2013) which lead to depersonalisation and people being cut off from themselves and others (Marcel, 1965) and distinguish it from a life centred on *being*, being present in the moment, and experiencing unity with other people and the world (Fromm, 2013; Marcel, 1965) through creating, loving, and sharing (Fromm, 2013).

This is not to imply that economic activity should be abandoned and, after we dismantle the market system, we return to a gatherer's life in caves. It is simply a matter of métron. Almost 300 years ago, Jeremy Bentham noticed that the satisfaction we get from the second cup of coffee is less than the satisfaction we get from the first; the law of diminishing marginal utility (Crimmins, 2021), which has been empirically confirmed in multiple studies (Kahneman & Tversky, 1979; Scitovsky, 1992; Tversky & Kahneman, 1981). "[H]aving too many choices produces psychological distress" (Schwartz, 2009) and leads to irrational choices that do not increase the overall value and satisfaction of the decision-maker (Kahneman & Tversky, 1979; Tversky & Kahneman, 1981). People are happy within métron but feel confused and diminished when it is exceeded. Offering the correct number of alternative products by taking into account the fact that the curve of the relationship between wealth or choices and happiness has an inverted U-shape[1] (Taleb, 2020) is the métron of success of the Future Manager.

The Future Manager, who has a regenerated mindset aiming for sustainable prosperity for all, uses data and technology to determine the optimal offering; that which maximises the value for the customers, reduces costs for the organisation, and minimises waste. This contributes to the organisation's profitability and long-term viability as well as helps control negative externalities. One Future Manager at a time and one organisation at a time will ultimately eliminate the negative externalities of exchanges. Until we get to the desired state of business working in tandem with institutions and society to bring about sustainable prosperity for all, Future Managers will be the opinion leaders encouraging others to embrace Transmodernism, learn from the world's great traditions, and exercise métron while acting locally but thinking globally.

This latter is particularly important as the externalities are not localised around the exchange episodes and, of course, involve many more parties than those directly, indirectly, or remotely involved. Negative externalities produced in local markets spread far and wide. This is neither a characteristic of globalisation and its progress nor restricted to the hyperconnected twenty-first century; a careless or badly timed intervention in one part of the system might cause serious, and often fatal, externalities in a seemingly unrelated part of the system at a time when it

---

[1] Simply put, this means that having no choice at all makes a person unhappy. Each choice or acquisition reduces this person's discomfort and creates a feeling of euphoria. There is, however, a point of maximisation of euphoria (the top of the curve) and from that point on, each subsequent acquisition or choice causes a decrease rather than an increase in well-being to the point where the number of choices or pile of possessions is so large that it makes the person unhappy again.

is least expected. Proof of this is that the rabbits released in the Australian bush 200 years ago are now pushing the Spanish eagles to extinction.

Irrespective of where one thinks we are on the hyper globalisation—deglobalisation continuum (Gong et al., 2022), how fast, or in which direction one thinks we are moving on it, the fact remains that we are, and have been for centuries now, inhabitants of a global village which is expanding over, above, and under the surface of the earth and consists of resources and actors inexorably linked through multiple dynamic networks. These dynamic networks comprise the global production networks (GPNs) and the global value chains (GVCs) with their communication media, norms, communicative products, interactions, relationships, atmospheres, and contextual frameworks through value cocreating exchanges take place (Cunningham, 1980; Ford, 1990; Gong et al., 2022; Hakansson, 1982). Information and communication technologies (ICTs), automations, transportation and logistics systems, but also geopolitical power shifts reconfigure GNPs and GVCs fast, often violently, and at an unprecedented scale. But the faster and easier people, materials, goods, services, and money move, the more the externalities spread and affect billions of people and a multitude of natural habitats and the species they host. Even though in recent years there have been numerous instances of both legal and covert barriers to the flow of people, goods, and funds between countries and regions, the essence of the global village does not change. We are all inexorably interconnected.

3.4.1 Reductionism Is No Longer Unavoidable; Enter Technology-Supported Holism

The multiplicity and interconnectedness of the twenty-first-century business milieu cannot be understood and addressed by twentieth-century tools such as the "PEST-type analyses" (Vlados & Chatzinikolaou, 2019) used for the macroenvironment, the diamond of national advantage for the mesoenvironmment (Porter, 1990), the 5 forces model for the competitive environment (Porter, 2008), and the 3BL model for sustainability. Such approaches compartmentalise the factors affecting each situation into predetermined categories, thus presenting a fragmented picture. Moreover, the categories are arbitrary and mask their interconnections and overlaps. Take any article on Trump's tariffs, for example, "*Trump's tariffs can remain in place for now, but appeals court fast tracks a summer resolution*" (Maruf & Cole, 2025), put yourself in the position of a Chinese or Indian manager in the steel industry and try to categorise the intelligence it provides into the factors that might potentially affect your firm under the categories of political, economic, legal, demand, factor conditions, rivalry, and threat of new entrants. The picture is so fragmented and fluid that it becomes blurred, like a constantly shaken and never allowed to set kaleidoscope.

The traditional analysis tools are as good as they can be. They offer decision-makers static two-dimensional puzzle pieces. Some of these pieces fit together. The ones that do not are ignored. The results of pieces of the puzzle being ignored are simply accepted as negative externalities to be borne by stakeholders who are neither part of nor party to the decisions. A fragmented and incomplete picture of reality, however, results in a messy analysis which, in turn, results in poorly supported decisions and, consequently, unpredictable results and potentially severe consequences for both overlooked stakeholders but also for the organisation. These consequences are simply accepted by practitioners as unavoidable risks which cannot be adequately managed and by academics as the generally acceptable limitations of the reductionist approach to decision-making.

Within the framework of modernity, the interplay between structures—such as the institutional (e.g. dominance of the sovereign nation-state model of political organisation) and socioeconomic (e.g. the distinction between developing and developed countries or the various class systems with their rules of membership and mobility) ones—and technology, with its capabilities and limitations (e.g. access to data, statistical methods, and computing power) practically mandated a reductionist approach. Reductionism prevailed for centuries in both science and philosophy from Democritus and Aristotle through Newton and Descartes to twentieth-century economics, biology, psychology, physics, and ecology, to name but a few scientific fields, and led to an understanding of complex systems as an agglomeration of their constituents. Reductionism reduces a complex system into its parts, reduces the parts of the system into their simplest possible form, regards them as independent, and categorises them in the smaller possible number of categories, and reduces the phenomena the system produces into the simplest possible behaviours and most direct possible interactions of the components. Finally, reductionism leads to seeking grand and unifying theories that can encompass the maximum number of phenomena and be applied to the maximum number of systems (Cat, 2024; El Hassani & Azdimousa, 2024; Popan, 2024). Hence the 'PEST-type analyses' and statistical analyses of country-level data which produce results like "a democratic environment can per se facilitate entrepreneurship and promote its growth" (Farè et al., 2023), or survey-based equations of human behaviour leading to statements such as "in-store … posters can influence as much as 70 percent of unplanned buying decisions" (Zhou & Wong, 2004) or "Gen Z considers transgressions on the part of macro influencers to be more serious than those of micro influencers; hence, they penalize macro influencers more severely" (Pradhan et al., 2023).

Philosophers of science have been challenging reductionism since the 1930s (Popper, 1935/[2005]) and their calls have been increasingly heeded by contemporary business scholars (El Hassani & Azdimousa, 2024; Hodgson, 2025; Jackson, 2016; Murmann et al., 2003; Vlados & Chatzinikolaou, 2019) who simply echo the lived experience of managers trying to navigate the permacrisis: "the current business environment, characterized by distinct mutations, complexity, and dynamism necessitates a fresh perspective" (El Hassani & Azdimousa, 2024). This fresh perspective is a departure from reductionism and an embracement of holism, which puts "the study of wholes before that of the parts" (Jackson, 2016).

Under the holism logic, systems (e.g. organisations, markets, industries, or counties) are viewed as coherent wholes which need to be understood so that we can ensure "that the parts are functioning and are related properly together so that they serve the purposes of the whole" (Jackson, 2016). Socioeconomic systems are by nature open systems surviving by interacting with the other systems which comprise their environment. Without constant exchanges they die; through repeated mutually satisfying exchanges they thrive. They can only thrive when that other systems in their environment also thrive and continue to engage in exchanges. Ultimately, serving their purpose depends on the other systems' serving theirs. They are all interconnected. *We* are all inexorably interconnected.

As the externalities are mounting in scope, reach, and severity (e.g. global warming, widening of inequalities, and stagflationary pressures in most Western economies), no system is immune of its own negative externalities. Moreover, in the current era of accessible knowledge and technology (such as big data, advanced analytics, robotics, supercomputing, nanotechnology, blockchain, machine learning, and AI), the methodological limitations of modernity can no longer serve as a valid excuse for accepting both the risks of poorly supported decisions and the negative externalities. The environmental forces shaping the twenty-first century business decision-making terrain constitute a "dynamically unified and evolutionary ... open and interactive ... systemic web" comprising the macro-meso-micro spanning "co-evolving spheres" of (a) spatially organised local, national, and transnational institutions with the socioeconomic structures they produce, (b) value-coproducing organisational functions which reflect the industrial dynamics and technological capabilities of their locale and time, and (c) the government interventions which shape, reflect, and reproduce the system (Vlados & Chatzinikolaou, 2019).

To view systems in their entirety, embrace their diversity, and understand their interactions requires critical, creative and design thinking, can be supported by sophisticated data analysis tools, and can lead to value coproduction for all without the negative externalities. We now have the knowledge and tools to perform the complex analyses required to mobilise societal, public but also privately controlled resources—both the operand (the tangible, static, and typically passive classical economic resources such as land, raw materials, machines, etc., which are acted upon to produce an effect) and operant (the intangible, dynamic, and active resources such as knowledge, routines, relationships etc., which act upon other resources and transform them into value-bearing entities) (Vargo & Lusch, 2004) ones—in a way that ensures that (a) value is maximised and enjoyed equitably by those directly, indirectly, or even remotely involved and affected and (b) negative externalities are immediately minimised and ultimately eliminated. Technology allows us to do it. Whether we do it or not is a matter of choice.

3.5 Values as the Guiding Principles of Management

We can choose to focus all our efforts on ensuring the biggest benefit for the largest number of stakeholder while refraining from any violation of the rights of minorities or any exercise of power over those weaker than us or the underprivileged.

We can choose values as our guiding principles. We can substitute an amoralistic pursuit of individual interests with a métron of what is right and wrong, enough and too much.

Values are the building blocks of morality and the subject of the field of philosophy that is known as ethics. We will resist the temptation to launch into a discussion of what values are and how they can be evaluated, arranged in hierarchies, espoused, and acted upon. There are excellent accounts of the problematisation over morality (Gensler, 1996; Hinman, 1994; Mackie, 1990; Scruton, 2012; Williams, 1985), and one more would be out of place in a discussion of Future Management. We will also resist the temptation to discuss the subjectivity, objectivity, or relativity of values. Suffice to say that, from the perspective of contractual theories, such as Rawls', values are the "stateable principles" representing relatively homogeneous, consensual, and authoritative beliefs which serve the purpose of peacefully resolving conflict (Williams, 1985) between people, organisations, or systems and within the decision-making unit. They delineate the constraints on the activities of "active, intelligent, participants in a partly competitive life" which are necessary for the balancing of their well-being with the "benefit for all the competitors" as well as with the benefit for classes with "morally valid claims to consideration" but no voice, such as the unborn, the marginalised, or the non-human (Mackie, 1990).

Values are the métron of appropriateness of the intentions behind a decision, the means employed to implement it, the ends it produces, and also the virtue of the actors—both the decision-makers and the other people involved in making and implementing a decision. Values guide the efficient and effective utilisation of operand and operant resources and the distribution of the value produced by this utilisation. Ultimately, it is values that give an organisation not only its moral legitimacy, which guarantees it the social acceptance required to be allowed to operate (Suchman, 1995), but also its fundamental reason for existing. Vision and mission statements, as well as explicit statements of organisational values, are part and parcel of organisational design for decades now. In the current troubled times and the turbulent ones ahead of us, it will be values that will hold organisations, communities, and societies together, and it is amoralistic profiteering that propagates cruelty and social disintegration.

There is, of course, a substantial body of literature on business ethics (Crane & Matten, 2007; De George, 1995; Dubbink & van der Deijl, 2023; Ferrell & Fraedrich, 1991; Ferrell & Fraedrich, 1997; Hoffman et al., 1994; Solomon, 1994; Velasquez, 1982; White, 1993), some of it focusing on describing what is done and some focusing on what should and should not be done and, increasingly, trying to balance a normative stance with the realities of the capitalist system. The key areas around which the bulk of the literature is concentrated are:

- The identification of stakeholders and how concern for their rights and allocation of value and externalities affects decision-making.
- The relationship between firms and their workers with respect to
  - industrial relations,

  - compensation,
  - meaningful work, and
  - whistleblowing.
- The role of corporations in society with regard to their
  - responsibilities
  - political activities
  - influence on and by cultures, especially in the context of transnational corporations and multinational firms.
- Marketing (and its alter ego, Purchasing) (Ford, 1990), that is, the interface between organisations and significant stakeholders outside the organisation.

The latter, being the organisational function most visible to society, law enforcement agencies, and policy makers, has attracted a lot of attention by scholars and has developed into a distinct field of scholarly inquiry (Eagle & Dahl, 2015; Eagle et al., 2020; Smith & Murphy, 2012; Smith & Quelch, 1992). Most of it is organised along the lines of the 4Ps and pinpoints problems in terms of product safety, pricing, promotion tactics, especially advertising, and distribution, with an emphasis on the ethics of withholding necessary goods or services from underprivileged consumers and segments which are assessed as being of low customer value. This approach, like all reductionist analyses, masks the interoperability of all business functions. Moreover, the distinctions between functions, agents, and stakeholders (e.g. like the one in the list of key areas above or the ones evident in the tables of contents of both philosophical (Moriarty, 2021) and business-oriented texts (McDonald, 2014)) blur the boundaries of responsibility and thus encourage managerial amoralism.

It has been strongly argued that the prevailing "economic order has now come to a point where the consequences of its amoralism are too deep and painful to ignore" (Panigyrakis & Zarkada, 2014). The late twentieth-century illusions of empowerment of the individual and affluence of the masses, global security and rule of international law, world order and tolerance of the incommensurable (Lyotard, 1984) have been shattered; its grand narratives, such as liberation, equality and progress (Lyotard, 1984; van Raaij, 1993), ring hollow in the ears of managers facing the permacrisis, and so, the need to re-examine our fundamental values is more evident than ever. And with 'our' we mean that we need to take stock of what makes each and every one of us human and humane, useful to those close to us, our community and the world, and apply our values to each and every decision we make as the simplest and safest way to ensure that we are agents of value coproduction. By assuming responsibility for the equitable distribution of value and the control of the negative social and environmental externalities, we can counter the burgeoning cruelty and curb the rising of despair around us. Conversely, it has been suggested (Panigyrakis & Zarkada, 2014) that the purpose of business is to mobilise and organise the operand and operant resources required for value cocreation in a manner that makes every person involved in the process a better human being and the world a better place.

3.5.1 Practical Applications of Values in Future Management

There are no invisible hands managing markets. It is people who make decisions that form all resource and product markets. People with faces and names who are accountable for their decisions and actions and bear the responsibilities for the consequences of their actions. Moreover, markets are not bipolar structures in which buyers possessed by unmet needs and unfulfillable desires meet profit-seeking sellers. They are complex systems comprising a variety of agents and affected parties, some of which bear the externalities without enjoying their fair share of value. This imbalance needs to be addressed, and the only way it can be addressed is by values-driven actions. Indeed, what unites all the scholarly and practitioner-oriented works on business ethics is the quest for the values which can serve as the foundation and connective tissue of individuals, organisations, markets, industries, and overseeing institutions.

Far be it from us to list our values. However, we would like to offer the following tools Future Managers can use to solve the moral dilemmas they face in the course of their professional life and to lead their organisations to lead the way to sustainable prosperity for all.

- Acknowledge the "primordial phenomenon of gentleness", the natural tendency of human beings to assume 'responsibility for the Other' (Levinas, 1980, p. 150), exercise métron over "rampantly egotistical desires", and use a hierarchy of values to make axiological judgements over all available courses of action (Panigyrakis & Zarkada, 2014).
- Bravely liberate oneself from the shackles of an engineered reductionist view of one's corporate role and adopt a holistic view of the self as a responsible agent and the organisation as a system and not as a structure comprising distinct subphenomena (the human experiences and sensory systems), its phenomena (relationships and networks), and its superphenomena (sustainability and development) (Achrol & Kotler, 2012).
- Seek the meaning of an individual life through truth and in truthfulness (Williams, 1985), or in contemporary business terminology emphasise the sequence of Trust  Commitment  Diligence (Murphy & Lasczniak, 1981), which, together with the assumption of responsibility for the totality of the social consequences of business decisions, is the bedrock of corporate sustainability (Gunelius, 2012).
- Instead of amoralistic marketing practice axiological societing; the summary rejection of arbitrary and unrealistic distinctions between product and packaging or price, "consumer and producer, customer and marketer as business executives and investors are also consumers and all of them are citizens of the same, one global village" (Panigyrakis & Zarkada, 2014). In practice, this means:
  - Instead of thinking in reductionist terms, which lead to outdated and impractical conceptualisations such as the 'dead' (Batat, 2022) traditional marketing mix, adopt a holistic approach, which emerges as the answers to four fundamental questions (Wichmann et al., 2022):

who is involved in the exchange function?—expect more parties to be more closely involved and the impact of brand, patent, or process and technology protectionism to be decreased
what constitutes the exchange system?—simplicity is likely to become the métron of success while added complexity is likely to work against consumer sentiment and to reduce profitability
how is the system of exchanges implemented?—automation will need to be balanced with the need of people to experience 'the human touch'
where are exchanges deployed?—the standardisation *vs* adaptation conundrum will become increasingly pressing while the physical and digital will merge further into a seamless phygital domain.

It follows that, even though the above points are elaborations on ideas which first appeared in marketing texts, their applicability is wider and, in the spirit of a holistic theory of management, can be used to describe any kind of exchange such as employment relations, resource acquisition, etc.

– Apply the consumer sovereignty test (Smith, 1995), which we propose can be extended to cover any party to an exchange, to determine if the who, what, how, and where of management practice is consistent with the values that were chosen as the métron of moral acceptability. In practical terms, this means that any value proposition or initiation and management of any exchange relationship needs to be assessed *vis-à-vis* the other party's:
  Capability to make free and informed decisions. In the presence of any vulnerability factors (as is the case with underage consumers, borrowers with low education or unstable income, or a job applicant with a mental impairment), the more capable of the two parties bears the responsibility of the weaker party's negative externalities. If people overborrow, it is the bank's fault.
  Access to sufficient information of appropriate quality. Very long and full-of-legalese insurance or telco contracts are a prime example of violations of the consumer sovereignty test. Too much or too little information which creates unrealistic consumer expectations not only leads to dissatisfaction but could also result in financial or emotional hardship for the poorly informed party.
  Choice. Monopolies, oligopolies, and cartels are direct violations of consumer sovereignty as they eliminate or limit choices and, usually, increase switching costs and thus violate the freedom to choose who to enter into a relationship with.

- Practice métron and use the tools of communication at their disposal to encourage others to also practice métron.

When values act as behaviour drivers, the world moves towards sustainable prosperity for all. When values are excluded from business decision-making, people find themselves in a state of discord with the organisations they enter into exchange relationships with. This is demoralising and counterproductive, as it erodes social capital, the "connections among individuals—social networks and the norms of reciprocity and trustworthiness that arise from them" (Putnam, 2001). When managers exercise their personal virtue, social capital, a form of civic virtue (Putnam, 2001), grows and flourishes, thus enriching both people's private lives and society as a whole. Virtuous behaviour encourages reciprocity—not the specific, *quid pro quo* type of obligation, but the social norm of generalised reciprocity which makes people do good and show kindness without expecting anything in return but believing in good things happening for them at another time, by a third party. This creates a culture of goodness, spreads virtue, and enriches a society with fine-grained, valuable social capital. Moreover,

> [w]hen economic and political dealing is embedded in dense networks of social interaction, incentives for opportunism and malfeasance are reduced. This is why the diamond trade, with its extreme possibilities for fraud, is concentrated within close-knit ethnic enclaves. Dense social ties facilitate gossip and other valuable ways of cultivating reputation—an essential foundation for trust in a complex society. (Putnam, 2001)

This chapter has presented a vision and a method for values-driven value cocreation without negative externalities. It is a projection of a positive outcome that virtuous managers can bring about despite the increasing severity and extend (both temporally and spatially) of the permacrisis. There are fragments of Future Management that are emerging into the current reality. One of them is the ocean parley shoes[2] and sports apparel which are made from marine plastic litter, one of the worst environmental problems. When all the oceans have been cleaned, we will have controlled the negative externalities we have been generating for years; when all shoes are made from marine litter, we will be on the way to environmental sustainability. When such products are news no more, then the business will have succeeded in implementing values-driven value cocreation. To do this, they will need to rethink, reimagine, and redesign their business model and operations. This process, strategic agility, is the subject of the next chapter.

---

[2] You can see details of the concept and products in the website of a sports company here: https://www.adidas.com/us/go/campaign/impact/planet and you can read reports from the specialist press (e.g. fashion https://ww.fashionnetwork.com/news/Adidas-and-parley-launch-3-new-tennis-shoe-designs-made-with-sustainable-materials,820936.html, retail https://www.retaildive.com/news/adidas-rolls-out-3d-printed-shoe-made-from-recycled-ocean-waste/429833/, design https://www.designboom.com/design/adidas-parley-shoe-concept-07-02-2015/) and business https://www.businessinsider.com/adidas-parley-for-the-oceans-special-sneakers-2016-6 and https://www.forbes.com/sites/afdhelaziz/2018/10/29/the-power-of-purpose-how-adidas-will-make-1-billion-helping-solve-the-problem-of-ocean-plastic/) as well as the general, mainstream news media https://www.reuters.com/article/business/adidas-to-launch-new-fabrics-from-recycled-ocean-plastic-polyester-idUSKBN1ZR1VQ/.

Future Management in Practice: Value and Values

This chapter's focus has been on providing an *Ethical Compass.* Future Management highlights that a focus on value alone, without the accompanying values, is inherently flawed and will not stand the test of time. This chapter illustrates how concentrating solely on the value of investors, employees, or customers leads to imbalances and negative externalities with detrimental effects for the managers, the organisation, society, and the world. For managers, this signifies that value creation is no longer a mere financial transaction but rather must be a product of ethical decisions that take into account future legitimacy.

1. Redefining Value

Value to managers goes beyond revenue or returns to shareholders. Future Management compels them to look at value in terms of an organisation's ability to improve the well-being of its employees, customers, communities and even the supporting ecosystems. The management function thereby undergoes a transformation; it is no longer a grasping for value but a giving in pursuit of greater good.

2. Securing the Values in the Decision-Making Process

It is essential to understand that values are neither supplementary nor public relations jingles. Rather, they serve as the guidepost for management decisions especially during conflicts. This entails a seamless blend of fairness, inclusivity, and transparency in strategy formulation, hiring, customer service, and supplier bargaining. Future Management demands that managers inquire, "what creates value?" but also "for whom, at what cost, and under what ethical conditions?".

3. Managerial Responsibilities in Practice

Having redefined prosperity as collective in Chapter 2, it is now the duty of managers to ensure that value is created and shared in a just manner. This involves the design of performance evaluation systems and incentive structures, and the development of partnerships that prevent exploitative practises and instead promote the cocreation of value among all stakeholders.

4. From Transactional to Relational Management

Future Management advocates for the elimination of transactional dealings in favour of trust-based relationships that are nurtured over the long term. Employees must be considered as purpose-driven partners, customers as co-creators, and communities as interested parties whose well-being is tied to organisational legitimacy. Such a shift demands investment in conversations, joint learning, and strategy adjustments in line with shared values.

In Practice:
Future Management means understanding that value and values cannot be separated. Managers who place values at the centre of strategy will not only reduce risks and externalities but will also create resilient and trusted organisations that can deliver sustainable prosperity.

MANAGERS MUST INTEGRATE VALUES INTO DECISION-MAKING, ENSURE FAIR VALUE CREATION, AND MOVE FROM TRANSACTIONS TO RELATIONSHIPS.

References

Achrol, R. S., & Kotler, P. (2012). Frontiers of the marketing paradigm in the third millennium. *Journal of the Academy of Marketing Science, 40*(1), 35–52.

Agle, B. R., & Caldwell, C. B. (1999). Understanding research on values in business. *Business and Society, 38*(3), 326–387.

Anderson, W. T. (1995). *The truth about the truth: De-confusing and re-constructing the postmodern world*. Jeremy Tarcher/Putnam.

Arnould, E. J., & Thompson, C. J. (2005). Consumer culture theory (CCT): Twenty years of research. *Journal of Consumer Research, 31*(4), 868–882.

Batat, W. (2022). Why is the traditional marketing mix dead? Towards the "experiential marketing mix"(7E), a strategic framework for business experience design in the phygital age. *Journal of Strategic Marketing*, 1–13.

Benson, J. (1937). Advertising looks ahead. *Journal of Marketing, 1*(3), 241–246. https://doi.org/10.2307/1245650

Billore, S., & Anisimova, T. (2021). Panic buying research: A systematic literature review and future research agenda. *International Journal of Consumer Studies, 45*(4), 777–804. https://doi.org/10.1111/ijcs.12669

Bonazzi, M. (2024). Protagoras. In E. N. Zalta & U. Nodelman (Eds.), *The Stanford encyclopedia of philosophy* (Fall 2024 ed.). Stanford University.

Boukis, A. (2019). Internal market orientation as a value creation mechanism. *Journal of Services Marketing, 33*(2), 233–244. https://doi.org/10.1108/JSM-10-2017-0336

Boukis, A., & Kabadayi, S. (2020). A classification of resources for employee-based value creation and a future research agenda. *European Management Journal, 38*(6), 863–873.

Bourdieu, P. (1984). *Distinction: A social critique of the judgement of taste*. Harvard University Press.

Bu, Y., Parkinson, J., & Thaichon, P. (2022). Influencer marketing: Homophily, customer value co-creation behaviour and purchase intention. *Journal of Retailing and Consumer Services, 66*, 102904. https://doi.org/10.1016/j.jretconser.2021.102904

Cat, J. (2024). The unity of science. In E. N. Zalta & U. Nodelman (Eds.), *The Stanford encyclopedia of philosophy* (Summer ed.). Stanford University.

Chapleo, C. (2010). Branding a university: Adding real value or 'smoke and mirrors'? *The Marketisation of Higher Education, 101*.

Chauhan, V. (2024). Barriers to adoption of slow tourism: An innovation resistance theory perspective. *International Journal of Tourism Research, 26*(4). https://doi.org/10.1002/jtr.2689

Che, Y., Zhu, J., & Huang, H. (2022). How does employee–organization relationship affect work engagement and work well-being of knowledge-based employees? *Frontiers in Psychology, 13*, Article 814324.

Churchill, W. (1935). Scientific pricing. *American Marketing Journal, 2*(1), 31–34.

Coffin, J., & Egan-Wyer, C. (2022). The ethical consumption cap and mean market morality. *Marketing Theory, 22*(1), 105–123.

Crane, A., & Matten, D. (2007). *Business ethics* (2nd ed.). Oxford University Press.

Crimmins, J. E. (2021). Jeremy Bentham. In E. N. Zalta (Ed.), *The Stanford encyclopedia of philosophy* (Winter ed.). Metaphysics Research Lab, Stanford University.

Crockett, D. (2021). Racial oppression and racial projects in consumer markets: A racial formation theory approach. *Journal of Consumer Research, 49*(1), 1–24. https://doi.org/10.1093/jcr/ucab050

Crossley, A. M. (1937). Measuring public opinion. *The Journal of Marketing, 1*(3), 272–274.

Csikszentmihalyi, M. (1999). If we are so rich, why aren't we happy? *American Psychologist, 54*(10), 821.

Cunningham, M. T. (1980). International marketing and purchasing of industrial goods—Features of a European research project. *European Journal of Marketing, 14*(5/6), 322–339.

Daryanto, A., de Ruyter, K., & Wetzels, M. (2010). Getting a discount or sharing the cost: The influence of regulatory fit on consumer response to service pricing schemes. *Journal of Service Research, 13*(2), 153–167. https://doi.org/10.1177/1094670509351566

De George, R. T. (1995). *Business ethics* (4th ed.). Macmillan.

Doyle, P. (2009). *Value-based marketing: Marketing strategies for corporate growth and shareholder value.* John Wiley & Sons.

Dubbink, W., & van der Deijl, W. (Eds.). (2023). *Business ethics: A philosophical introduction.* Springer.

Dwivedi, Y. K., Ismagilova, E., Hughes, D. L., Carlson, J., Filieri, R., Jacobson, J., Jain, V., Karjaluoto, H., Kefi, H., Krishen, A. S., Kumar, V., Rahman, M. M., Raman, R., Rauschnabel, P. A., Rowley, J., Salo, J., Tran, G. A., & Wang, Y. (2021). Setting the future of digital and social media marketing research: Perspectives and research propositions. *International Journal of Information Management, 59*, 102168. https://doi.org/10.1016/j.ijinfomgt.2020.102168

Eagle, L., & Dahl, S. (2015). *Marketing ethics & society.* SAGE Publications.

Eagle, L., Dahl, S., De Pelsmacker, P., & Taylor, C. R. (2020). *The SAGE handbook of marketing ethics.* SAGE Reference.

Edvardsson, B., Tronvoll, B., & Gruber, T. (2011). Expanding understanding of service exchange and value co-creation: A social construction approach. *Journal of the Academy of Marketing Science, 39*(2), 327–339. https://doi.org/10.1007/s11747-010-0200-y

El Hassani, M., & Azdimousa, H. (2024). Beyond traditional marketing: Holistic marketing as the key to success in the era of the 4th industrial revolution and Post-Covid. In M. Ezziyyani, J. Kacprzyk, & V. E. Balas (Eds.), *International Conference on Advanced Intelligent Systems for Sustainable Development (AI2SD'2023)* Cham.

Farè, L., Audretsch, D. B., & Dejardin, M. (2023). Does democracy foster entrepreneurship? *Small Business Economics, 61*(4), 1461–1495. https://doi.org/10.1007/s11187-023-00737-7

Ferrell, O. C., & Fraedrich, J. (1991). *Business ethics.* Houghton Mifflin Company.

Ferrell, O. C., & Fraedrich, J. (1997). *Business ethics: Ethical decision making and cases.* Houghton Mufflin Co.

Floridi, L. (2018). Semantic capital: Its nature, value, and curation. *Philosophy & Technology, 31*(4), 481–497. https://doi.org/10.1007/s13347-018-0335-1

Ford, D. (Ed.). (1990). *Understanding business markets: Interaction, relationships and networks.* Academic Press.

Fotiadis, T., Folinas, D., Vasileiou, K., & Konstantoglou, A. (2022). *Marketing and the customer value chain.* Taylor and Francis.

Fromm, E. (2013). *To have or to be?* Bloomsbury Academic.

Galbraith, J. K. (1998/[1958]). *The affluent society* (40th annivesary ed.). Houghton Mifflin Harcourt.

Gasana, P. U. (2009). Relative status and interdependent effects in consumer behavior. *The Journal of Socio-Economics, 38*(1), 52–59. https://doi.org/10.1016/j.socec.2008.07.009

Gelles, D., & Schleifer, T. (2025). Climate Group funded by Bill Gates slashes staff in major retreat. *The New York Times*. Retrieved April 3, 2025, from https://www.nytimes.com/2025/03/12/climate/bill-gates-breakthrough-energy-cuts.html#:~:text=Breakthrough%20Energy%2C%20an%20umbrella%20group,resetting%20for%20the%20Trump%20era.&text=Breakthrough%20Energy%2C%20an%20umbrella%20organization,an%20internal%20memo%20on%20Tuesday

Gensler, H. J. (1996). *Formal ethics*. Routledge.

Glaveli, N., & Karassavidou, E. (2011). Exploring a possible route through which training affects organizational performance: The case of a Greek bank. *The International Journal of Human Resource Management, 22*(14), 2892–2923. https://doi.org/10.1080/09585192.2011.606113

Gong, H., Hassink, R., Foster, C., Hess, M., & Garretsen, H. (2022). Globalisation in reverse? Reconfiguring the geographies of value chains and production networks. In (Vol. 15, pp. 165–181), Oxford University Press UK.

Granot, E., Alejandro, T. B., & Russell, L. T. M. (2014). A socio-marketing analysis of the concept of cute and its consumer culture implications. *Journal of Consumer Culture, 14*(1), 66–87.

Grönroos, C. (2011). A service perspective on business relationships: The value creation, interaction and marketing interface. *Industrial Marketing Management, 40*(2), 240–247. https://doi.org/10.1016/j.indmarman.2010.06.036

Grönroos, C., & Voima, P. (2013). Critical service logic: Making sense of value creation and co-creation. *Journal of the Academy of Marketing Science, 41*(2), 133–150. https://doi.org/10.1007/s11747-012-0308-3

Gunelius, S. (2012). 10 brand marketing trends that should dominate 2013. *Forbeswoman*. http://www.forbes.com/sites/work-in-progress/2012/12/18/10-brand-marketing-trends-that-should-dominate-2013/

Gupta, A. (2021). Definitions. In E. N. Zalta (Ed.), *The Stanford encyclopedia of philosophy* (Winter ed.). Stanford, CA: Stanford University.

Hakansson, H. (Ed.). (1982). *International marketing and purchasing of industrial goods*. John Wiley & Sons.

Healy, J. C., & McDonagh, P. (2013). Consumer roles in brand culture and value co-creation in virtual communities. *Journal of Business Research, 66*(9), 1528–1540.

Hess, H. W. (1935). The new consumption era. *American Marketing Journal, 2*(1), 16–25.

Hinman, L. M. (1994). *Ethics: A pluralistic approach to moral theory*. Harcourt Brace College Publishers.

Hodgson, G. M. (2025). *From Marx to markets: An intellectual Odyssey*. Edward Elgar Publishing.

Hoffman, W., Kamm , M., Frederick, J., Robert, E., & Petry, J., E. S. (1994). *Emerging global business ethics*. Westport.

Holbrook, M. B. (1987). What is consumer research? *The Journal of Consumer Research, 14*(1), 128–132.

The Holy Qur'an: Translation. (2000). (A. Y. Ali, Trans.; Vol. Tahrike Tarsile Qur'an). Wordsworth.

Hungara, A., & Nobre, H. (2021). A consumer culture theory perspective of the marketplace: An integrative review and agenda for research. *International Journal of Consumer Studies, 45*(4), 805–823.

Iglesias, O., Ind, N., & Schultz, M. (2022). Towards a paradigm shift in corporate branding. In O. Iglesias, N. Ind, & M. Schultz (Eds.), *The Routledge companion to corporate branding*. Routledge.

Ingenbleek, P. (2007). Value-informed pricing in its organizational context: Literature review, conceptual framework, and directions for future research. *Journal of Product & Brand Management, 16*(7), 441–458.

Jackson, M. C. (2016). *Systems thinking: Creative holism for managers*. Citeseer.

Kahneman, D., & Tversky, A. (1979). Prospect Theory: An analysis of decisions under risk. *Econometrica, 47*(2), 313–327.

Kashif, M., & Zarkada, A. K. (2015). Value co-destruction between customers and frontline employees: A social system perspective. *International Journal of Bank of Marketing, 33*(6), 672–691. https://doi.org/10.1108/IJBM-09-2014-0121
Kelly, S., Johnston, P., & Danheiser, S. (2017). *Value-ology*. Palgrave Macmillan.
Kokkinopoulou, E., Papasolomou, I., Vrontis, D., & Porcu, L. (2024, 2024). Co-creating sustainable buying behavior among Generation Z consumers: The role of earned media. Marketing solutions to the challenges of a VUCA Environment, Cham.
Kotler, P., Pfoertsch, W., & Sponholz, U. (2020). *H2H Marketing: The genesis of human-to-human marketing*. Springer International Publishing.
Lee, C. H., & Bruvold, N. T. (2003). Creating value for employees: Investment in employee development. *The International Journal of Human Resource Management, 14*(6), 981–1000.
Levinas, E. (1980). *Totality and infinity: An essay on exteriority* (Vol. 1). Kluwer.
Levine, M. (2007). Challenging the culture of affluence. *Independent School, 67*(1), 28–36.
Levy, S. J. (1959). Symbols for sale. *Harvard Business Review*(July–August), 117–124.
Löbler, H. (2013). Service-dominant networks: An evolution from the service-dominant logic perspective. *Journal of Service Management, 24*(4), 4–4.
Locke, J. (1690). *An essay concerning humane understanding*. Thomas Basset.
Lumivalo, J., Tuunanen, T., & Salo, M. (2024). Value co-destruction: A conceptual review and future research agenda. *Journal of Service Research, 27*(2), 159–176. https://doi.org/10.1177/10946705231177504
Lusch, R. F., & Vargo, S. L. (2006). Service-dominant logic: Reactions, reflections and refinements. *Marketing Theory, 6*(3), 281–288. https://doi.org/10.1177/1470593106066781
Lyotard, J.-F. (1984). *The postmodern condition: A report on knowledge (La Condition postmoderne: rapport sur le savoir)* (G. Bennington & B. Massumi, Trans.). University of Minneapolis Press.
MacIntyre, A. C. (2007). *After virtue* (3rd ed., Vol. 211). University of Notre Dame Press.
Mackie, J. L. (1990). *Ethics: Inventing right and wrong*. Penguin.
Mahajan, G. (2017a). *How creating customer value makes you a great executive*. Business Expert Press.
Mahajan, G. (2017b). Value dominant logic. *Journal of Creating Value, 3*(2), 217–235.
Mahajan, G., Kumar, V., Tregua, M., & Bruni, R. (2023). Value-dominant logic: Organizational principles. *European Journal of Marketing, 57*(5), 1245–1271. https://doi.org/10.1108/EJM-11-2022-0816
Mansfeld, J. (1981). Protagoras on epistemological obstacles and persons. In G. B. Kerferd (Ed.), *The sophists and their legacy* (pp. 38–53). Cambridge University Press.
Marcel, G. (1965). *Being and having: An existentialist diary*. Harper & Row.
Maruf, R., & Cole, D. (2025). Trump's tariffs can remain in place for now, but appeals court fast tracks a summer resolution. *CNN Business*. https://edition.cnn.com/2025/06/10/business/tariffs-appeals-court-stay-trump
Marx, K. (2018/[1867]). *The capital* (Vol. 1–3). Madison & Adams Press.
Maycroft, N. (2004). Cultural consumption and the myth of life-style. *Capital & Class, 28*(3), 61–75.
McCarthy, E. J. (1960). *Basic marketing: A managerial approach*. Irwin Inc.
McDonald, G. (2014). *Business ethics: A contemporary approach*. Cambridge University Press.
McVey, L., Gurrieri, L., & Tyler, M. (2021). The structural oppression of women by markets: The continuum of sexual violence and the online pornography market. *Journal of Marketing Management, 37*(1–2), 40–67. https://doi.org/10.1080/0267257X.2020.1798714
Mingione, M., & Leoni, L. (2020). Blurring B2C and B2B boundaries: Corporate brand value co-creation in B2B2C markets. *Journal of Marketing Management, 36*(1–2), 72–99.
Moriarty, J. (2021). Business ethics. In E. N. Zalta & U. Nodelman (Eds.), *The Stanford encyclopedia of philosophy* (Fall ed.). Stanford University.
Morris, S., & Snell, S. (2019). How employees can better solve customer problems: a use value approach to human and social capital. In A. J. Nyberg & T. P. Moliterno (Eds.), *Handbook of research on strategic human capital resources*. https://doi.org/10.4337/9781788116695.00023

Muniz, F., & Guzmán, F. (2023). The impact of brand value co-creation on perceived CSR authenticity and brand equity. *Journal of Product & Brand Management, 32*(8), 1338–1354. https://doi.org/10.1108/jpbm-02-2023-4340

Murmann, J. P., Aldrich, H. E., Levinthal, D., & Winter, S. G. (2003). Evolutionary thought in management and organization theory at the beginning of the New Millennium: A symposium on the state of the art and opportunities for future research. *Journal of Management Inquiry, 12*(1), 22–40. https://doi.org/10.1177/1056492602250516

Murphy, P., & Lasczniak, G. R. (1981). Marketing ethics: A review with implications for managers, educators and researchers. *Review of Marketing*, 251–266.

Nash, B. (1937). Product development. *The Journal of Marketing, 1*(3), 254–262.

Panigyrakis, G. G., & Zarkada, A. K. (2014). New philosophical paradigms in marketing: From amoral consumerism to axiological societing. In L. Moutinho, E. Bigne, & A. K. Manrai (Eds.), *The Routledge companion on the future of marketing*. Routledge.

Parasuraman, A., Zeithaml, V. A., & Berry, L. L. (1985). A conceptual model of service quality and its implications for future research. *Journal of Marketing, 49*(4), 41–50. https://doi.org/10.2307/1251430

Parsons, E., Kearney, T., Surman, E., Cappellini, B., Moffat, S., Harman, V., & Scheurenbrand, K. (2021). Who really cares? Introducing an 'Ethics of Care'to debates on transformative value co-creation. *Journal of Business Research, 122*, 794–804.

Paul, J., & Rosenbaum, M. (2020). Retailing and consumer services at a tipping point: New conceptual frameworks and theoretical models. *Journal of Retailing and Consumer Services, 54*, 101977. https://doi.org/10.1016/j.jretconser.2019.101977

Payne, A. F., Storbacka, K., & Frow, P. (2008). Managing the co-creation of value. *Journal of the Academy of Marketing Science, 36*(1), 83–96.

Pelsmacker, P., Janssens, W., & Mielants, C. (2005). Consumer values and fair-trade beliefs, attitudes and buying behaviour. *International Review on Public and Non Profit Marketing, 2*(2), 50–69. https://doi.org/10.1007/bf02893257

Petrovic, G. (1991). Reification. In T. Bottomore, L. Harris, V. G. Kiernan, & R. Miliband (Eds.), *A dictionary of marxist thought* (2nd ed., pp. 463–465). Blackwell.

Plato. (2008). *Protagoras* (B. Jowett, Ed.). Project Gutenberg.

Plé, L., & Cáceres, R. C. (2010). Not always co-creation: Introducing interactional co-destruction of value in service-dominant logic. *Journal of Services Marketing, 24*(6), 430–437.

Pokorny, M. (2013). Getting to Know your employees and what motivates them. *Employment Relations Today, 39*(4), 45–52. https://doi.org/10.1002/ert.21389

Popan, E. (2024). Reductionism (philosophy). *Research Starters*. https://www.ebsco.com/research-starters/religion-and-philosophy/reductionism-philosophy#:~:text=Reductionism%20in%20philosophy%20is%20the,explained%20by%20their%20simpler%20parts.

Popper, K. R. (1935/[2005]). *The logic of scientific discovery*. Taylor & Francis e-Library.

Porter, M. (1990). *The competitive advantage of nations*. The Free Press.

Porter, M. E. (2008). The five competitive forces that shape strategy. *Harvard Business Review, January*, 23–41.

Pradhan, D., Kuanr, A., Anupurba Pahi, S., & Akram, M. S. (2023). Influencer marketing: When and why gen Z consumers avoid influencers and endorsed brands. *Psychology and Marketing, 40*(1), 27–47. https://doi.org/10.1002/mar.21749

Putnam, R. D. (2001). *Bowling Alone: The collapse and revival of American community* (Revised and Updated ed.). Simon & Schuster.

Roig, J. C. F., García, J. S., & Moliner, M. Á. M. (2009). Perceived value and customer loyalty in financial services. *The Service Industries Journal, 29*(6), 775–789. https://doi.org/10.1080/02642060902749286

Roig, J. C. F., Garcia, J. S., Tena, M. A. M., & Monzonis, J. L. (2006). Customer perceived value in banking services. *International Journal of Bank Marketing, 24*(5), 266–283. https://doi.org/10.1108/02652320610681729

Rosa, S. (1663–1664). *Democritus and Protagoras* [oil on canvas]. Hermitage Museum hall 238. https://commons.wikimedia.org/wiki/File:Salvator_Rosa_-_D%C3%A9mocrite_et_Protagoras.jpg

Rotaru, I., Nitulescu, L., & Rudolf, C. (2010). The post-modern paradigm—A framework of today's media impact in cultural space. *Procedia—Social and Behavioral Sciences, 5*, 328–330. https://doi.org/10.1016/j.sbspro.2010.07.098

Russell, B. (2004). *A history of western philosophy*. Routledge. (Original work published 1946)

Scanlan, J. P. (1967). Nikolaj Černyševskij and Soviet philosophy. *Studies in East European Thought, 7*(1), 1–27.

Schmitt, B., Brakus, J. J., & Biraglia, A. (2021). Consumption ideology. *Journal of Consumer Research, 49*(1), 74–95. https://doi.org/10.1093/jcr/ucab044

Schroeder, M. (2021). Value theory. In E. N. Zalta (Ed.), *The Stanford encyclopedia of philosophy* (Winter 2021 ed.). Metaphysics Research Lab CSLI Stanford University.

Schwartz, B. (2009). *The Paradox of choice: Why more is less*. HarperCollins e-books.

Scitovsky, T. (1992). *The joyless economy: The psychology of human satisfaction*. Oxford University Press on Demand.

Scruton, R. (2012). *Modern philosophy: An introduction and survey*. A&C Black.

Shankar, A., & Fitchett, J. A. (2002). Having, being and consumption. *Journal of Marketing Management, 18*(5–6), 501–516. https://doi.org/10.1362/0267257022683721

Shankar, A., Whittaker, J., & Fitchett, J. A. (2006). Heaven knows I'm miserable now. *Marketing Theory, 6*(4), 485–505. https://doi.org/10.1177/1470593106069933

Shaw, D., Grehan, E., Shiu, E., Hassan, L., & Thomson, J. (2005). An exploration of values in ethical consumer decision making. *Journal of Consumer Behaviour, 4*(3), 185–200. https://doi.org/10.1002/cb.3

Shein, E. (2025). Saying 'Please and Thank You' to ChatGPT is burning cash and energy: Why it matters. In *eWeek* (pp. N.PAG–N.PAG): TechnologyAdvice, LLC.

Simon, H. A. (1955). *A behavioral model of rational choice*. Oxford University Press.

Simon, H. A. (1979). Rational decision making in business organizations. *The American Economic Review, 69*(4), 493–513.

Smith, A. (1761). *The theory of moral sentiments* (2nd ed.). A. Millar & A. Kincaid and J. Bell.

Smith, N. C. (1995). Marketing strategies for the ethics era. *MIT Sloan Management Review*.

Smith, N. C., & Murphy, P. E. (2012). *Marketing ethics*. SAGE Publications.

Smith, N. C., & Quelch, J. A. (1992). *Ethics in marketing*. Irwin.

Solomon, R. C. (1994). *Above the bottom line: An introduction to business ethics* (2nd ed.). Harcourt Brace College Publishers.

Srivastava, R. (2019). *Decoding customer value at the bottom of the pyramid*. Business Expert Press.

Suchman, M. C. (1995). Managing legitimacy: Strategic and institutional approaches. *Academy of Management Review, 20*(July), 571–610. https://doi.org/10.5465/amr.1995.9508080331

Tadajewski, M., & Jones, D. G. B. (2011). Scientific marketing management and the emergence of the ethical marketing concept. *Journal of Marketing Management, 28*(1–2), 37–61. https://doi.org/10.1080/0267257x.2011.619072

Taleb, N. N. (2020). *Skin in the game: Hidden asymmetries in daily life*. Penguin e-book.

Teece, D. J. (2007). Explicating dynamic capabilities: The nature and microfoundations of (sustainable) enterprise performance. *Strategic Management Journal, 28*(13), 1319–1350. https://doi.org/10.1002/smj.640

Tversky, A., & Kahneman, D. (1981). The framing of decisions and the psychology of choice. *Science, 211*(4481), 453–458. https://doi.org/10.1126/science.7455683 (New Series)

Van Boven, L., & Gilovich, T. (2003). To do or to have? That is the question. *Journal of Personality and Social Psychology, 85*(6), 1193–1202. https://doi.org/10.1037/0022-3514.85.6.1193

van Raaij, W. F. (1993). Postmodern consumption. *Journal of Economic Psychology, 14*(3), 541–563. https://doi.org/10.1016/0167-4870(93)90032-g

van Waterschoot, W., & Van den Bulte, C. (1992). The 4P classification of the marketing mix revisited. *Journal of Marketing, 56*(4), 83–93. https://doi.org/10.1177/002224299205600407

Vargo, S. L. (2009). *Service-dominant logic: An alternative mindset for innovation.* 2009 Ministry of the Knowledge Economy Conference on Service Innovation through R&D, Seoul. http://www.sdlogic.net/Seoul_RD_2009.pdf

Vargo, S. L., & Lusch, R. F. (2004). Evolving to a new dominant logic for marketing. *Journal of Marketing, 68*(1), 1–17.

Vargo, S. L., & Lusch, R. F. (2011). It's all B2B…and beyond: Toward a systems perspective of the market. *Industrial Marketing Management*, *40*(2), 181–187. https://doi.org/10.1016/j.indmarman.2010.06.026

Vargo, S. L., & Lusch, R. F. (2018). *The SAGE handbook of service-dominant logic*. Sage.

Varul, M. Z. (2011). Value: Exchange and use value. In D. Southerton (Ed.), *Encyclopedia of consumer culture* (pp. 1501–1503). Sage.

Veblen, T. (1998/[1899]). *The theory of the leisure class: An economic study of institutions.* Prometheus Books.

Velasquez, M. (1982). *Business ethics.* Prentice-Hall.

Verhoef, P. C., Lemon, K. N., Parasuraman, A., Roggeveen, A., Tsiros, M., & Schlesinger, L. A. (2009). Customer experience creation: Determinants, dynamics and management strategies. *Journal of retailing*, *85*(1), 31–41. https://doi.org/10.1016/j.jretai.2008.11.001

Vlados, C., & Chatzinikolaou, D. (2019). Methodological redirections for an evolutionary approach of the external business environment. *Journal of Management & Sustainability, 9*(2), 25–46.

Webster, F. E., Jr. (1992). The changing role of marketing in the corporation. *Journal of Marketing, 56*(4), 1–17.

Weinstein, A. (2018). *Superior customer value* (4th ed.). Taylor and Francis.

Weinstein, A., & Johnson, W. C. (2020). *Designing and delivering superior customer value.* Taylor and Francis.

White, T. I. (1993). *Business ethics: A philosophical reader*. Macmillan Pub.

Wichmann, J. R. K., Uppal, A., Sharma, A., & Dekimpe, M. G. (2022). A global perspective on the marketing mix across time and space. *International Journal of Research in Marketing, 39*(2), 502–521. https://doi.org/10.1016/j.ijresmar.2021.09.001

Wilk, R. (1999). Towards a Useful Multigenic Theory of Consumption. European Council for an Energy Efficient Economy (ECEEE) 1999 Summer Study proceedings: Energy efficiency and CO2 reduction: the dimensions of the social challenge (web version).

Wilk, R. (2002). Consumption, human needs, and global environmental change. *Global Environmental Change, 12*(1), 5–13. https://doi.org/10.1016/S0959-3780(01)00028-0

Williams, B. (1985). *Ethics and the limits of philosophy.* Fontana Press.

Wirtz, J., & Lovelock, C. (2016). *Services marketing: People technology strategy* (8th ed.). World Scientific.

Woodall, T. (2012). Driven to excess? Linking calling, character and the (mis) behaviour of marketers. *Marketing Theory, 12*(2), 173–191.

Yi, Y. (2014). *Customer value creation behavior*. Taylor and Francis.

Zeithaml, V. A., Berry, L. L., & Parasuraman, A. (1993). The nature and determinants of customer expectations of service. *Journal of the Academy of Marketing Science, 21*(1), 1–12.

Zeithaml, V. A., Verleye, K., Hatak, I., Koller, M., & Zauner, A. (2020). Three decades of customer value research: Paradigmatic roots and future research avenues. *Journal of Service Research, 23*(4), 409–432.

Zhou, L., & Wong, A. (2004). Consumer impulse buying and in-store stimuli in Chinese supermarkets. *Journal of International Consumer Marketing, 16*(2), 37–53. https://doi.org/10.1300/j046v16n02_03

Αριστοτέλης (Aristotle). (*trans.* 2015/[4th century BCE]). *Ἀναλυτικὰ Ὕστερα (Posterior Analytics)* (G. R. G. Mure, Trans.). eBooks@Adelaide, The University of Adelaide Library.

Πλάτων (Plato). (2015/[369 BCE]). *Θεαίτητος: Η γνωσιολογία του Πλάτωνος (Theaetetus: Plato on Knowledge)* (Κ. Σ. Παπαλεξίου (K. S. Papalexiou), Trans.). Ζήτρος.

Designing and Managing for Strategic Agility

4

4.1 Introduction: The Need for Strategic Agility

In the current environment fraught with unprecedented turmoil and complexity, the capability of an organisation to anticipate, adapt, and thrive in highly fluid environments is more critical than ever. Strategic agility, the ability to adjust to changes in the competitive landscape while maintaining a focus on core purpose, has thus become a critical impermanent for businesses in today's environment of chaos. This chapter lays out the principles, frames, and practical applications of strategic agility to offer insight towards building resilience, innovation, and sustainability in the longer run.

If there was one single company that could be used as a case study to demonstrate a powerful example of strategic agility, that would be Toyota Motor Corporation. Formulated around the mid-twentieth century as an innovative approach to production, the Toyota Lean Production System changed the whole manufacturing sector by advocating for just-in-time production, continuous improvement (kaizen), and elimination of wastages. Unlike the traditional mass production systems, which depended on large inventories and inflexible processes, the Toyota lean production model made it possible for the company to quickly adjust itself according to changes in market demand, enhancing efficiency and minimising wastage at the same time.

Toyota consistently seeks to optimise its operations, eliminating any inefficiencies, thus securing their competitive advantage in the automotive high volatile industry. There have been numerous challenges for this company over several decades, including economic depressions, natural disasters and disruptions in supply chains, but Toyota has been able to overcome them by having a flexible and adaptive manufacturing process. It is this flexibility and adaptability that is necessary for operating in such an unpredictable business environment.

A. K. Zarkada and Y. Melanthiou, *Future Management for Sustainable Prosperity*, Future of Business and Finance, https://doi.org/10.1007/978-3-032-15868-0_4

These ideologies made Toyota a serious contender in the face of lean periods in economies, technological revolutions, and large-scale logistical crises. Toyota works on a macro-interpretation of the Dynamic Capabilities Theory (Teece, 2007), suggesting that companies that evolve, coordinate, and recombine their assets can sustain a competitive advantage. Toyota indeed epitomises operational agility, proving that organisations should actively design their firms to innovate and be flexible continuously. This chapter delineates a pathway through which organisations might realise agility in their structures, leadership, and decision-making frameworks to be able to keep pace with the rapidly changing environment of today.

Strategic agility is difficult to achieve, though not entirely impossible. Beyond operational efficiency, strategic agility is rooted in Complexity Theory in Management (Stacey, 1995), which acknowledges that organisations operate in unpredictable and nonlinear environments. Where traditional planning is static for the most part, complexity theory emphasises the need for adaptability, decentralised decision-making, and iterative learning. This aligns with the Organisational Learning Theory (Argyris & Schön, 1997), which underscores the importance of continuous feedback loops and adaptive learning processes in fostering agility.

In today's world catalysed by extensive technological advancements, the spectre of global unrest, and ever-changing consumer expectations, organisations may no longer depend on rigid, long-term strategies. A focus on strategic sensing, consumer-led updates, real-time decision-making, and innovation-driven resilience is their only hope for survival against the disruption. The concept of a "permacrisis", where instability is the norm rather than the exception, necessitates a shift from traditional strategic planning to an agile mindset that balances foresight with flexibility.

This chapter will explore key frameworks that support strategic agility, including environmental scanning models such as the traditional PESTLE framework but also emerging alternatives like the BANI (Brittle, Anxious, Nonlinear, Incomprehensible), the Three Horizons Model by McKinsey, and the Cynefin Framework. The objective of this chapter is to point out the relevant case studies and working business insights for any leader or decision-maker who truly desires the successful design and management of an agile organisation in its uncertain world.

4.2 Navigating the External Environment: PESTLE Analysis

For businesses to achieve strategic agility, they need to consider all outside forces that may impact their operations. For this wide-ranging environmental scanning, the PESTLE—Political, Economic, Sociocultural, Technological, Legal and Environmental—framework serves the purpose.

Political instability and policy change, such as trade laws, can produce severe business and supply chain implications, the effects of political instability. Brexit, implemented in 2021, significantly disrupted industries like construction, where restrictions on EU labour led to critical workforce shortages (UK Parliament, 2022). Organisations are re-evaluating sourcing strategies and markets for the sake of minimising risks, as in the case of Russia-Ukraine war, which has dramatically impacted the world trade and supply chains (World Bank, 2022). With Ukraine being one of the most significant exporters of grains and metals in the world, and Russia supplying crude oil and gas, it is clear that this war will have a broad economic effect (European Commission, 2023). Moreover, there have been acute shortages, price fluctuations, and disruptions in agriculture, manufacturing, and energy (International Monetary Fund, 2023). In response to this, global businesses diversified their supply chains, modified their purchasing strategies, and searched for other ways to do business to minimise their geopolitical risks (McKinsey & Company, 2023).

Historical events such as the Global Financial Crisis of 2007–2009 have made it abundantly clear that, in an environment of economic uncertainty, strategic agility would take precedence over the flexibility in an organisation's structural design. Companies that adapted quickly managed to survive, while rigid organisations struggled. More recently, the COVID-19 pandemic reiterated the importance of predictive analytics driven by AI and scenario planning; many corporations had to quickly reconfigure supply chains, workforce management, and digital strategies just to survive in unpredictable environments (Manyika & Bughin, 2018).

Transforming the dynamics of the market are shifting demographics, evolving cultural norms, and changing consumer preferences, and ensuring adaptability holds the key to business survival (Mintel, 2022). This means that companies should remain socially responsive and proactive in getting their products and services aligned with the current values of a society. One example is the shift towards plant-based diets by consumers who now demand more nutritious, environmentally friendly, and ethical food choices (Stoll-Kleemann & Schmidt, 2022), thus aiding in the mainstreaming of products like Beyond Meat and Oatly, and creating an impetus for Nestlé and McDonald's to grow their product lines adding plant-based ranges in order to capture this market (Euromonitor, 2021).

The fast-evolving technological advances ensure continuous disruptions in industries that require sustained investments in innovation, digital transformation, and cybersecurity to stay competitive (Bughin et al., 2021). The application of a traditional disruptive model by Uber by smartphone technology, GPS, and app-based ride-hailing has been the most notable example in the tech economy and app-based services' disruption of the traditional market—the model was later adopted by competitors like Bolt and Grab (McKinsey, 2022). Digitalisation has raised consumer expectations in terms of seamless online experience and personalised services, required by businesses to either enhance their digital presence or protect themselves from cybersecurity breaches.

Legal and regulatory factors are important even regarding business continuity and risk management. Evolving laws can severely affect operations, and non-compliance can become extremely challenging for businesses, often leading to financial penalties (World Economic Forum, 2022). Since its launch within the EU in 2018, GDPR has been influential in the shaping of global data privacy legislation, with more enforcement actions being taken in recent years. These have led to record fines against companies like Amazon and Meta (formerly Facebook)—$877 million against Amazon and $1.3 billion against Meta—for violations of data protection laws (European Data Protection Board, 2023). This means that businesses will, in all likelihood, need to keep shifting their compliance strategies as one of the avenues through which to mitigate financial risks and stay in the good books of consumers.

Environmental factors, and particularly climate change and sustainability issues, are becoming of paramount importance for business. Businesses are required to reduce gas emissions, switch to renewable sources of energy, and adopt sustainable practises. For instance, a Middle Eastern-based small-scale enterprise in the United Arab Emirates (UAE) called Beeah Group has become a leader in environmental innovation in the Middle East by operating the region's first waste-to-energy plant reducing landfill usage.

The PESTLE analysis is greatly held in assessing external environmental factors that may affect organisational strategies. Yet technological advancements, geopolitical shifts, and economic instability are evolving so rapidly that merely one-off environmental scanning is not enough to keep an organisation on the fence. With the pace of developments today, businesses must be able to qualitatively and quantitatively address realities of the external business environment in real time to anticipate and respond proactively to disruptions.

Aware of these developments, emerging strategic frameworks and technologies that can be integrated with traditional PESTLE analysis include BANI (Brittle, Anxious, Nonlinear, Incomprehensible), AI-facilitated predictive analytics, scenario planning 2.0, and the Cynefin framework. These approaches enhance decision-making by offering more adaptive, data-driven insights into external complexities.

4.2.1 Enhancing PESTLE for Strategic Agility: Beyond Traditional Frameworks

Although PESTLE provides an orderly method for scanning external environments, it does not sufficiently capture the increasingly complex and fragile state of contemporary systems. To enhance strategic foresight, organisations should integrate emerging new frameworks such as AI-based predictive analytics and BANI, rendering them more dynamic and proactive in their response to uncertainty.

AI-Driven Predictive Analytics and Scenario Planning 2.0

AI has moved PESTLE analysis from a static approach to an agile, real-time strategic tool. The past involved organisations identifying political, economic, social, technological, legal, and environmental problems by way of iterated organisational assessments, mostly leading to critical time lags in response to external interruptions. AI-driven analytics now allow firms to constantly monitor, predict, and respond much more accurately and quickly to external influences, ensuring a proactive strategy formulation rather than a reactive one to cut future losses.

One of the areas in which AI significantly boosts analysis in politics and economics is the machine learning algorithms, which scan oceans of global data to flag early signs of shifts in policy, changes to trade regulations, and trends in financial markets. Organisations can now start preparing for changes in regulations and reorganising their strategies instead of waiting until the actual effects of the legislation affect the organisation. AI applications combine and analyse news, government reports, and even feelings felt by policymakers to identify certain measurable patterns or outcomes in international trade policies by which central banks could make their decisions regarding interest rates and currencies. Such predictive intelligence allows organisations to employ risk-hedging strategies by which, for example, they can diversify their supply chains before tariffs or sanctions take effect.

Increasingly, AI impacts the social dimension of PESTLE, especially with consumer sentiment analysis. Taking intelligence from vast data samples, ranging from social media data, online reviews, and buying patterns, allows AI to prepare organisations for changes on the consumer side and in society. For example, AI can already sense new things about sustainability intentions so that companies can already start working on launching environmentally sustainable-known products just before competition notices it. This real-time analysis of everything social means that companies are thus able to increase their strategic agility by ensuring that the products and services they offer are in pace with ever-evolving societal values, thus enhancing brand perception and customer loyalty.

Just as with the technological and legal aspects, the impact of AI in those areas is equally drastic. In the technology area, AI facilitates real-time monitoring of emerging innovations, resulting in making businesses stay afloat when faced with industry disruption. AI use in patent analysis and forecasting research trends gives the company an early advantage of identifying breakthrough technologies before they become mainstream. Furthermore, the legal side will be related to enhancing compliance management use of AI through automated regulatory monitoring. As AI scans legal databases, court rulings, and industry regulations, it keeps businesses on the right side of the ever-changing laws, minimising penalties and business-busting interruptions.

AI is revolutionising environmental analysis, risk modelling in climate change, and sustainability forecasts. An organisation can use AI to analyse the long-term impacts of environmental factors such as changing climate, carbon emission regulations, and resource scarcity on its operations. By equipping itself with AI-powered risk assessment tools, organisations will actively review their sustainability strategies for alignment with global policies on the environment or investor expectations.

Furthermore, Scenario Planning 2.0 (Ramírez and Wilkinson, 2016) attempts to develop AI-driven simulations of future crises and strategic reactions as opposed to static forecasts. Contrary to traditional scenario planning that relies on human foresight, AI modelling refreshes continuously per real-time information about changing geopolitical, economic, or technological contexts.

For instance, here again, to an extent outweighed by economic outlook periodic reports, organisations like Amazon and Netflix follow AI in tracking shifts in customer behaviour towards instant pricing, inventory, and content management.

Embedding AI within PESTLE analysis propels entities from a passive method of agility in strategy toward proactivity: anticipating external threats, discerning emerging opportunities, and making informed, real-time decisions. This transformation equips companies with the confidence to navigate the choppy waters of uncertainty and keeps them strategically focused on long-term competitiveness in an increasingly volatile business context.

Beyond VUCA: The Emergence of the BANI Framework

While the traditional holding on conceptualising turbulence in the environment was done through the VUCA framework (Volatile, Uncertain, Complex, and Ambiguous), it misses equating fully the current nature of disruptions to the world created by fragile systems and layers of excessive information that induce a nonlinear chain reaction.

In this context, the BANI (Brittle, Anxious, Non-Linear, Incomprehensible) (Cascio et al, 2025) constitutes a more appropriate model for organisations navigating exponential change, digital transformation, and systemic fragility. BANI reflects the fragility of modern supply chains, the psychological cost of uncertainty, and the difference between technological change accelerating theoretically and exponentially. The PESTLE allows for a structural view of the external forces, and BANI allows one to interpret how those external forces operate in unpredictable and uncertain ways, which, in turn, means that businesses are required to adapt strategies that have resilience, modularity, and adaptability.

For instance, in a VUCA-based supply chain strategy, emphasis could be placed on supplier diversification for the mitigation of risks stemming from such geopolitical tensions. On the other hand, BANI-based supply chains prioritise decentralisation, real-time monitoring, and blockchain-driven transparency thus increasing their members' ability to adapt to sudden collapse and disruptions.

The Three Horizons Model and Cynefin Framework, Two Decision-Making Compasses in Complexity

Although PESTLE analyses external environments in a systematic way, it doesn't always give organisations clear directions on the actions to take or how to deal with complexity. There are two strategic frameworks that address such gaps:

1. The Three Horizons Model (McKinsey)

This model (Curry & Hodgson, 2008) shows that strategic planning involves three horizons so that organisations can manage their long-term vision with agility:

- Horizon 1 The short-term upsurge: Business as usual; day-to-day activities efficient and effective.
- Horizon 2 Mid-term innovation: Applying best developing opportunities to move from current business model to future.
- Horizon 3 Long term vision: Invest for new disruptive innovations.

Tesla, for example, operates across all three horizons—an optimised present EV production (H1), a scaled battery technology (H2), and an investment in AI-driven autonomous mobility (H3).

2. The Cynefin Framework: Classifying Complexity

The Cynefin framework (Snowden & Boone, 2007) provides a decision-making framework for withstanding some areas of uncertainty through the classification of problems into four types:

- Clear (Simple): well-established best practices apply (e.g. routine compliance regulations).
- Complicated: requiring expert analytical reasoning through predictable patterns of behaviour (e.g. a change in tax policies).
- Complex: no known solutions; need to experiment and understand (e.g. adapting to AI-driven market changes).
- Chaotic: instant action required; immediate steps needed to deal with a crisis (e.g. sudden geopolitical conflicts).

When combined with PESTLE knowledge, in relation to decisions arrived at using the Cynefin model, organisations would know which external trends need structured responses and which call for agile experimentation.

The PESTLE analysis will be there for environmental reporting, but its relevance in the fast pace of change will surely be enhanced by combining it with frameworks like BANI, predictive analytics powered by AI, scenario planning, and the Cynefin model. Hence, organisations can seamlessly transition from static environmental analysis to real-time sensing, adapt to and anticipate non-linear disturbances rather than merely react, try to find a balance between short-term agility and long-term strategic foresight, and adopt different decision-making models based on the complexity of external challenges. This paradigm shift from passive environmental scanning to active strategic sensing would ensure organisational survival during uncertain and dynamic terrains while also creating an environment for it to thrive.

4.3 Resilience, Agility, and Sustainability: A Triad for Success

The three most essential factors defining the success of long-term organisational effectiveness—resilience, agility, and sustainability—are becoming increasingly indispensable in an increasingly uncertain and volatile world. Businesses can overcome challenges, adjust to novel circumstances, and survive through time. Resilience makes it possible for an organisation to withstand unplanned events and to maintain continuity. Agility facilitates shifts required in response to such unplanned events but also seeks further opportunities. Sustainability is a reliability guarantee by ensuring that such actions taken align with the businesses' long-term objectives and societal values, thereby ensuring the foundation of long-term success (Fig. 4.1).

4.3.1 Understanding Resilience in a Business Environment

Business resilience is characterised by a critical capability that enables the survival of organisations amid uncertainty, crises, and disruptive change (Duchek, 2020). Put in other words, resilience refers to an organisation's ability to anticipate, absorb, and adapt to shocks with operational continuity. Thus, resilience is a key enabler of long-term agility and future competitiveness (Hamel & Valikangas, 2003). Resilient-strategized organisations not only tend to stay in operations while facing disruptions but also seem to evolve into stronger companies that innovate or transform from crises (Radovic-Markovic et al., 2023; Lengnick-Hall et al., 2011).

In a business world increasingly characterised by volatility, it is no longer a matter of bouncing back from disruptions; it is about continuously adapting to and flourishing in uncertainty. Resilience embedded proactively in strategies not just prepares organizations to tide through crises but also proactively adjust to changes in market conditions and, eventually, position themselves to exploit emerging opportunities as they come along (Duchek, 2020). However, resilience is not the same in every organisation; it manifests in different dimensions depending on how a business anticipates and reacts to risks. Attaining comprehensive

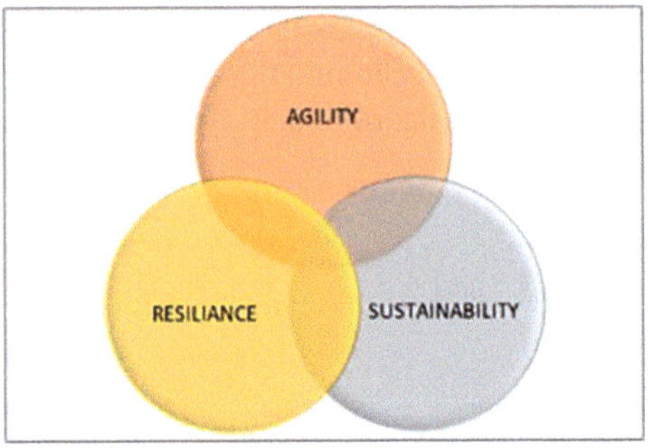

Fig. 4.1 Overlap between resilience, agility and sustainability. Image created by Yioula Melanthiou

resilience entails organising the three areas of focus in resilience, which are operational, strategic, and financial (Hamel & Valikangas, 2003; Reeves & Whitaker, 2020), all of which tie into the long-term stability and agility of organisations.

Types of Organisational Resilience

Operational Resilience: Ensures continuity of business functions by making changes in supply chains, workforce structures, and types of digital infrastructures to meet external shocks (Sheffi, 2015). For example, large-scale changes in operations may have put companies like Amazon and Zoom into a different category during coronavirus-induced demands, while most other companies were not able to deliver (McKinsey & Company, 2021).

Case Study: Zoom's Response to the Remote Work Trend

zoom

When COVID-19 led to lockdowns all over the globe, Zoom was quick to respond by making good use of the situation to sell remote solutions. This tech company matched the increased demand for virtual meet-ups by increasing its operations and at the same time making its platform more secure and user-friendly. In adapting so swiftly, Zoom strengthened its market position and redefined its business model.

Traditional retailers like Debenhams in the UK, had difficulties adjusting during the pandemic. Having already been late to develop an effective e-commerce model, and with social distancing and quarantines in place not allowing for physical shops to remain open, forced the company to eventually close all its physical stores after over 200 years in operation.

Strategic Resilience: Extending beyond recovery, focuses on reinvention and sustainable competitiveness (Reeves & Whitaker, 2020). Businesses like Netflix that have fully transformed their operations from one form of business to another—from rental to streaming—affirm how continuous adaptation to market trends fosters resilience in quickly evolving landscapes (Girotra & Netessine, 2014).

Financial resilience: Maintaining strong financial buffers, revenue diversification, and risk management strategies to resist economic shocks (Barney, 2021). Firms that were scenario planners and liquidity managers ahead of the 2008 Financial Crisis were better positioned to survive its aftermath (Van der Vegt et al., 2015).

Building a Resilient Organisation

Resilience should, in effect, be built on a foundation of proactive investigation and a systems approach that integrates risk intelligence, workforce adaptability, and digital transformation for safeguarding long-term stability in uncertain times (Boin et al., 2021). Resilient organisations embed resilience within their core strategies by fostering continuous learning and strategic foresight, and by absorbing and recovering from shocks with operational effectiveness (Duchek, 2020). Cultivating a corporate culture that fosters adaptability is thus central to resilience. Those employees who embrace a growth mindset and maintain flexibility in responding to change elevate the organisation's ability to pivot during crises while remaining productive. Culturally resilient organisations promote leadership development, cross-functional collaboration, and scenario-based decision-making to build a workforce that is ready to embrace uncertainty (Sutcliffe & Vogus, 2003).

Technology is key for resilience. Organisations that employ AI and predictive analytics will be better prepared to cope with disturbances, allowing for the formulation of contingency plans before crises unfold (McKinsey Global Institute, 2022). In practice, AI-powered risk assessment tools help organisations analyse real-time data to identify weak signals indicating potential disruption sooner and with greater accuracy, thereby enabling organisations to adapt strategies in real time. Technologies like digital twins and simulation modelling allow companies to stress-test various crisis scenarios and optimise response plans to ensure their preparedness against future challenges (Reeves & Whitaker, 2020). Digital transformation also provides strategic advantages to improve supply chain resilience due to better visibility, greater automation, and decentralised decision-making. By integrating digital infrastructure into their supply chains, businesses can rapidly reconfigure sourcing and distribution networks to mitigate exposure to geopolitical or economic shocks (Sheffi, 2015).

Besides technological advances, financial resilience is crucial for sustaining an organisation through economic downturns and industry disruptions. Companies with diversified revenue streams, strong liquidity buffers, and good risk management strategies are in a better position to weather any financial shocks (Barney, 2021). The 2008 financial crisis proved that companies with stringent financial planning and models for scenario-based forecasting could navigate through economic turbulence compared to firms that relied on static budgeting practices (Van der Vegt et al., 2015). Hence, it is the future organisations that introduce stress testing and dynamic capital allocation models into their financial mechanism that will find it persuasive to evolve according to unforeseen changes in the economy (Tallon, 2021).

Resilience is thus enhanced by supply chain diversification and ecosystem partnerships. Stakeholders that share a singular source of supply are highly susceptible to disruption, as seen in the far-reaching shortages from the COVID-19 pandemic (McKinsey & Company, 2021). In mitigation of the risks posed by dependence on global supply chains, companies can multi-source, near-shore operations, and form collaborative networks with regional suppliers (Christopher & Peck, 2004). Companies like Tesla control key production processes through vertical integration to facilitate supply chain flexibility and reduce reliance on external vendors (Reeves & Whitaker, 2020).

Thus, resilient organisations do more than react to crises; they continuously adjust their strategies to anticipate and navigate uncertainty. Their cultural embrace of change, technological innovation, financial fortitude, and operational dependence diversification give businesses a competitive edge amid increased turbulence. Such turbulence is a disturbing characteristic of the future economy; those organisations that infuse resilience into their DNA will not merely survive the onslaught of disruption but will emerge as agile players in future challenges (Coutu, 2002; Hamel & Valikangas, 2003).

Resilience as Competitive Advantage

In an era where economic shocks, cyber threats, and climate crises are increasingly common, resilience is not just about survival—it is a competitive differentiator (Tallon, 2021). Organisations that focus on resilience are more likely to grab opportunities that emerge out of such uncertainty, thus fuelling their future growth and market leadership (Coutu, 2002). Successful companies need to ensure that there are systems, processes, and cultures in place capable to adapt to change without comprising operation continuity. A resilient organisation is one that continues to manage at least core functions in time of crises, be they economic, supply chain, or geopolitical in nature. One important aspect of resilience is the capacity to learn from disruption and to improve systems in the future to prevent becoming vulnerable. Organisations that invest in managing risks, preparing contingency plans, and empowering the workforce are likely to recover more quickly from setbacks and emerge stronger.

Case Study: Netflix Strategic Agility

Netflix success is built on its ability to recognize and proactively respond to changes in the external environment that have reshaped the entertainment industry over the past two decades. Initially founded in 1997, Netflix was a DVD rental service provider. It had foreseen that physical media would soon be a thing of the past and taking advantage of the fast-spreading broadband internet services and the rising demand for personalized TV shows, it boldly transitioned to streaming services long before any of its competitors. This revolution in technology alongside habits such as binge-watching programs online, meant that the company would have an early mover advantage within a new industry landscape. They soon seized another opportunity amidst mounting rivalry among streaming entities by entering the content production segment. By producing its original contents like House of Cards or Stranger Things, Netflix was able to own its content catalogue thus gaining a competitive advantage through product differentiation. Unlike Netflix,

Blockbuster once a top renter in video tapes, failed to transition following the disruptions in the entertainment industry, and ceased operations in 2014.

Agility as a Core Business Competency

Agility is something much more than just rapid response to change—it is a structured capability that allows a business to anticipate, innovate, and evolve in unpredictable environments (Bronlet, 2023). While resilience ensures an organisation survives crises, agility is that which ensures the effective utilisation of change for competitive advantage. Organisations that embed agility in their strategic planning nurture dynamic capabilities for continuous model enhancement, market repositioning and new technologies integration (McKinsey & Company, 2023). Unlike the traditional business models that rely on rigid cycles of planning, agile organisations shifted to iterative decision-making and real-time strategic shifts based on changing economic and technological markets.

Type(s) of Agility in Business Strategy

Operational Agility: Adapting Core Processes for Efficiency

Operational agility is the ability of an organisation to reconfigure, restructure, or rebuild its systems, supply chain, and workflows promptly in response to external disruptions (Reeves & Whitaker, 2020). Companies like Amazon and Walmart show they have operational agility in place by integrating the utilisation of AI in inventory management using predictive analytics to facilitate on-demand logistics (McKinsey Global Institute, 2022). An inflexible operational structure proves extremely problematic during unexpected conditions; for example, in the wake of COVID-19, traditional retailers failed due to supply chain breakdowns.

Strategic Agility: Rapidly Repositioning the Business Model

Strategic agility is the ability of remodelling an existing business model when markets shift, customer tastes change, or competitive threats arise. Netflix, for example, abandoned its DVD rental service and developed an AI-driven streaming platform, contributing to continuous improvement of its content strategy based on customer data (Girotra & Netessine, 2014). Microsoft implemented its own successful transformation from a software provider to a cloud-first business, precipitated by an Azure execution that has enabled it to fight effectively in the changing tech industry.

Cultural Agility: Embedding Flexibility in Leadership and Workforce

Agile companies foster cultural flexibility, from top management to front-line employees. Decentralised decision-making and cross-functional collaboration are prioritised, and the willingness for embracing calculated risks is prioritized. Agile practice at Spotify, for instance, involves forming small 'squads' that can independently check and suggest new product ideas without slowing down the bulk of company operations. Companies cultivating this cultural agility empower employees to respond dynamically to change, keep on learning, and continuously experiment.

Agility as a Competitive Advantage

As more industries continue to experience extraordinary volatility, agility will not just be an option in organisations but will be a main competitive determinant. Building agility into operational, strategic, and cultural dimensions ensures that a company not only has better weapons with which to handle its part in the market but also proves to be the better equipped to become a leader in changing times. Agility permits companies to actively shape rather than react to industry trends and hence becomes a key part of a sustainable competitive advantage in today's business landscape (Tallon, 2021).

The Role of Sustainability in Strategic Planning

Sustainability has progressed from corporate social responsibility to being a strategic mandate driving organisational resilience and competitiveness in the market. Improving long-term growth, companies that emphasise sustainability end up lowering risks associated with environmental and regulatory changes as well as creating new revenue streams and greater brand loyalty (Sancak, 2023). Neglecting to integrate sustainability into the business model places companies under the threat of regulatory action, resistance from consumers, and decreased investor confidence (European Commission, 2023).

Three Pillars of Sustainable Business Strategy

1. Environmental Sustainability: Reducing Carbon Footprint and Reducing Resource Dependence

Businesses are following directives of regulators, investors, and customers, moving towards carbon–neutral operations and circular economy models. Large companies like Unilever or IKEA propose future carbon reduction targets while implementing closed-loop production systems that reduce waste and extend product life (McKinsey & Company, 2023). Renewable energy adoption, sustainable sourcing, and climate risk adaptation have become fundamental in defining long-term business strategies, ensuring companies fit within future ecological developments.

2. Social Sustainability: Changing Consumer and Workforce Motivations

Consumers show greater interest in brands that confirm their ethical values, and brands exhibiting indifference within this trend are threatened to lose market share. For example, Patagonia embeds environmental activism into the identity of the brand itself, allocating generous profits into sustainability initiatives and financially supporting socially responsible consumerism (Tallon, 2021). Additionally, companies need workplace sustainability that ensures diversity, equity and inclusion, in order to attract and retain the top talent required in the competitive labour market.

3. Economic Sustainability: Sustainable Business Models as Foundations of Future Long-Term Economic Growth

Sustainability is a key business imperative, not only a matter of ethical concern but also as a long-term strategy for good financial performance. Firms adopting a sustainable supply chain, good labour practice, and ESG (Environmental, Social, and Governance) framework are particularly attractive to investors and stakeholders. Companies like Tesla and Amazon demonstrate how sustainable innovation processes in electric mobility and carbon-efficient logistics can boost overall financial success quite dramatically and protect against emerging risks.

Sustainable companies are intrinsically more agile because they have built a business based on long-term value rather than quick profits. By integrating sustainability with strategic planning, companies can anticipate changing regulations and shifting consumer preferences, and ensure business resilience. As industries are operating within sustainability-driven economies, organisations that consider sustainability as fundamental to their core strategies will emerge as leaders in the next wave of global business transformation (European Data Protection Board, 2023).

Case Study: Ikea's Circular Economy Approach

One company that has embedded sustainability into its core strategy is IKEA. The furniture giant has made extensive investments to ensure the operationally derived renewable energy generation makes it climate-positive by 2030. IKEA's sustainability strategies include sourcing 100% of its wood from sustainable sources, reducing waste through circular economy practices, and providing repair, recycling, and repurposing services. Prominent among its programs, is a commitment to the circular economy; IKEA aims to develop its products so that they last longer and are recyclable. This reduces waste by extending the cycles of its products. The introduction of furniture lease programs involves renting furniture instead of buying, which is another major step taken by the company in promoting sustainable practices. Certainly, one cannot overlook the fact that sustainability brings good reputation and customer loyalty to the brand. Putting sustainability deep into the very business model, IKEA has demonstrated that responsible business practises can drive innovation, growth, and long-term viability.

4.4 Purpose-Driven Companies: Agility Through Shared Values

Purpose-driven companies can better deal with the uncertainties and changing aspects of market conditions. A clearly defined purpose allows organisations to align perfectly with the demanding expectations from various stakeholders—customers, employees, investors, and society at large. Purpose serves as a guiding principle to ensure that even if companies pivot towards external strategic challenges, they do not lose sight of their core values (Henderson & Van den Steen, 2015). Such companies tend to exhibit higher levels of resilience and agility, and high trust from stakeholders.

4.4.1 The Role of Purpose in Driving Strategic Agility

Purpose provides clarity and direction in a volatile, uncertain, complex, and ambiguous (VUCA) environment. A well-connected and trusted organisation aids improved acceleration and strategic whirling in an altering external environment—the core tenets of strategic agility (Williams et al., 2021). Companies with a well-defined purpose are proactive in anticipating risks and tapping into new opportunities. Purpose-oriented companies align their actions to their mission, meaning that any business decision will substantiate their values and contribute to the greater good. Companies such as Alibaba, focusing on sustainability and social accountability, are in a better position to respond to changes in regulatory changes, consumer preferences, or environmental challenges. By prioritising long-term impact gains over short-term impact dividends, purpose-driven organisations can create conditions for growth while still safekeeping the trust of the stakeholders.

Case Study: Alibaba's Digital Philanthropy and Inclusive Growth

Alibaba is years ahead in addressing social inequality through digital philanthropy or through initiatives for inclusive growth. For example, an initiative in 2014 that launched Rural Taobao was meant to close the digital gap between urban and rural China. The centers opened in rural villages would enable local entrepreneurs to assist residents with online product selling or digitization literacy through e-commerce platforms. It enabled farmers to connect directly with urban consumers while eliminating middlemen and selling at better prices. For example, farmers in Jinzhai County earn more because they know how to sell their specialty tea products to more expanded markets. This is a really great example of empowering underserved communities through digital tools, thus promoting social equity along with economic growth. Beyond the Rural Taobao, the global SME Empowerment Program established by Alibaba gives small and medium enterprises digital tools and training to digitize and scale operations globally. On-demand and sustainable production for small manufacturers by means of cloud computing and IoT is promoted by Xunxi Digital Factory. These well-known digital inclusion initiatives boost the brand reputation of Alibaba because it is in line with its longer-term goals. Thus, by promoting inclusive growth, it expands its markets while creating customer loyalty and making its business future-proof. Alibaba demonstrates that agility can create winners in terms of profits as well as impact.

4.4.2 Establishing a Purpose-Driven Organisation

Establishing a purpose-driven organisation requires a systematic approach to integrating a company's mission into every aspect of the operations of the organisation. A well-defined purpose is the starting point. This involves coming up with a mission statement that defines key values and future directions for the company, laying the groundwork for making decisions and strategizing on future goals. According to the research, a well-articulated corporate objective is the foundation for keeping your business on the right course through making decisions guided along with the alignment of internal systems towards long-term organisational objectives (Bartlett & Ghoshal, 1994).

Once the purpose is defined clearly, it must be translated into the organisation's strategy. Business strategy, operations, and decision-making systems must reflect and support the mission, making sure consistency and coherence are evident throughout all levels. Unified direction creates strong internal grounding with

Fig. 4.2 Practical steps to build a purpose-driven organisation. Image created by Yioula Melanthiou

external perception of the brand. A purpose-driven organisation is one that actively engages its stakeholders. The organisation must communicate its mission to all employees, customers, investors, and communities that can enable it to achieve the goals set for that organisation. Finally, stakeholder engagement will help build trust, cooperation, and shared accountability, which are necessary for long-term success.

Measurement of impact, or proof of impact, is important for ensuring sustainability for the purpose-driven organisation. Employing metrics to gauge progress against social and environmental goals ensures accountability and transparency. The data-driven approach allows organisations to align with purpose and be adequately informed in adjusting their strategies. Continuous adaptability from the organisation will also identify newly arising problems and untapped opportunities that may potentially be explored. The dynamics of the business world require continued assessments of the mission and its execution. Purpose-rooted operations thus provide companies with the opportunity for sustainable growth alongside their commitments towards both social and environmental values (Fig. 4.2).

4.5 Authenticity and Trust: Building Agile, Transparent Organisations

Authenticity has become a distinguishing feature of organisations in the current dynamic context of business, striving to become competitive and remain relevant. Authentic organisations are the ones that are operably transparent and have value-based congruency between their values and actions. Such authenticity creates stakeholder trust with end users, employees, investors, and communities and allows the organisation to adapt quickly to external changes. In a world characterised by ambiguity and complexity, trust is still a strategic asset that can help businesses navigate disruptions and embrace new opportunities.

4.5.1 The Importance of Authenticity in Future Organisations

Authenticity is a moral imperative, but it also makes sense from a business perspective. It enhances stakeholder trust and aligns internal values with external actions, thereby advancing organisational legitimacy and long-term success (Avolio & Gardner, 2005). Authentic organisations often have loyal clients since they align their values and actions with their target audience's expected values and desires. Today, consumers require organisations to be open and honest, and any organisation that has authenticity built into its system is better positioned to have and maintain its loyalty from customers.

Authenticity also plays a significant role in employee commitment and satisfaction. Employees tend to see higher purpose and belonging when they perceive that their organisations are true to their values and act with integrity. This will manifest in higher productivity, lower turnover, and a more pleasant workplace culture. According to research by Ilies et al. (2005), perceiving an authentic leader and organisation provides the employees with an opportunity for working with increased engagement levels, psychological safety, and job satisfaction. Authentic organisations are also better placed to develop capable supply chains and partnerships. Doing business with ethical processes and transparent relationships will mitigate risks and create efficacious value chains through resilient supply chains. Resilient supply chains are the initial provisions required to handle business interruptions.

Finally, authenticity provides a competitive advantage and improves brand image. Companies that show authenticity in their words and deeds build the trust of their customers, employees, and stakeholders. Increasingly, in the current business environment, trust is what is worth having, and an authentic brand can set itself apart from its competitors and gain permanence in its success amidst the continuously changing business landscape.

4.5.2 Trust as a Strategic Asset

Trust is essential for organisational agility and resilience. Stakeholders are much more likely to support a company's initiatives for change and take on challenges during hard times. All stakeholders are more inclined to innovate and take on risks in an environment that values trust and authenticity. Even customers trust their favourite brands enough to continue buying from them even when the economy falls into recession.

In times of uncertainty and ambiguity, trust can become a distinct strategic asset that can provide a competitive advantage to organisations. Research has shown that in time of a crisis, trust reduces the need for excessive controls while speeding up decision-making and improving stakeholder cooperation (Bachmann & Inkpen, 2011). It comes from consistent and transparent communication, ethical leadership, and concern for stakeholder well-being. Trust reduces reliance on uncertainty and nurtures a positive culture within the organisation, wherein employees and stakeholders feel appreciated and understood.

Trust-oriented organisations are much more agile and resilient; they take decisions more quickly due to low resistance from stakeholders and respond swiftly and effectively to external changes in the market demand. We also find empirical evidence wherein trust turns into a gift in enabling psychological safety, risk-taking, and knowledge-sharing in formal and informal teams or networks (Dirks & Ferrin, 2001).

Case Study: Patagonia's Authentic Leadership

Patagonia is a good example to demonstrate organisational authenticity. The company donates its profits to environmentally friendly causes, fully aligning with their purpose. It is for this purpose and actions that Patagonia has a very loyal group of customers, who share similar principles and values regarding social and environmental concerns. In 2022, Patagonia's founder, Yvon Chouinard, bestowed the ownership of the company to a trust and a non-profit organisation which focuses on fighting climate change. This decision reinforced the image of authenticity by showing the unparalleled commitment of the company to its mission. This integrity and transparency further strengthened the trust of the stakeholders and positioned Patagonia as leader in its market.

4.5.3 Building an Authentic Organisation

Building an authentic organisation necessitates intentionally and strategically embedding key values into every facet of the organisation. The first step is to itemise and clarify what the core values of the company are. These values should not be statement-like or abstract, but rather concrete principles that direct the organisation's everyday operations. Take, for example, Patagonia, which demonstrates a full embodiment of sustainability through its business model as well as its practices. As such, defining core values lays the groundwork for an authentic organisational culture.

Once core values have been established, it then becomes mandatory that actions would align with these values. Business practices, policies, and communications need to consistently reflect throughout what the company is all about. According to Gardner et al. (2005), organisational authenticities are facilitated by the presence of a consistent alignment between held values and enacted behaviours, in particular when those behaviours are carried out by every leader with an ethical bearing. A dedicated concern for sustainability, for instance, would result in ensuring sustainable sourcing policies within a supply chain. This harmony between what is a core value and what really happens builds trust and credibility among stakeholders.

Another aspect of authenticity is transparency. A company often must communicate its objectives, challenges, and achievements. Reports on progress and developments can be shared, and businesses should be willing to demonstrate accountability and foster trust. Transparency assures stakeholders that the company is indeed genuinely committed to its values.

Fostering ethical leadership is of paramount importance for an authentic organisation. Ethical leadership extends beyond rules; it consists of doing what is right, even when that is not the easiest path. Being an ethical leader, modelling ethical behaviour, and focusing on ethical leadership ensure a business develops a culture of integrity.

Finally, stakeholders must be engaged in the decision-making process. By listening to the concerns and feedback of stakeholders, an organisation is in better standing to develop long-lasting relationships and have committed decisions that serve the needs and interests of those for whom they are committed to delivering value. Companies can hold Town Halls for example and invite stakeholders to gather feedback or concerns regarding various initiatives in place. This openness and willingness to listen and consider views shows respect and the willingness to learn and adapt if necessary (Kernis & Goldman, 2006) (Fig. 4.3).

Fig. 4.3 Practical steps to build an authentic organisation. Image created by Yioula Melanthiou

4.6 Agility in Action: Balancing Planning and Experimentation

Strategic agility is the hallmark of balancing long-term planning with short-term adaptability. For this reason, comprehensive plans make it difficult for organisations to meet unplanned disruptions because traditional management mainly emphasises detailed and strict planning. On the contrary, agility requires that organisations emphasise more flexibility through the application of experiments to adjust their operations while still pursuing their long-term direction. Such organisations exploit a plethora of opportunities in the market and formally mould them as per their intentions while consuming changes in market trends. The idea of organisational ambidexterity thinks of firms as simultaneously engaging themselves in exploitation (of their existing capabilities) and exploration (finding new opportunities) to be agile and competitive (O'Reilly & Tushman, 2013). Within a disruptive and uncertain business environment, this conceptual difference in structuring formalised planning with ad hoc adaptability is urgent. The orientation to only long-term planning will eventually lead to stagnation, and the full concentration on short-term reactions might not give any direction and coherence. A synergised approach in both cases creates a shock absorber and a push for innovative and sustainable strides in the company's growth.

Case Study: Amazon's Innovation Process

amazon

Another good balance between long-term planning and short-term adaptability can be found in Amazon, which excels in both innovation and adaptation abilities. Amazon is established through an obsession with customer needs, as well as a strong spirit of experimentation. They considered the innovation mode of Amazon by "working backward." This simply means their approach starts with the outcome that they want the customer to have and works back from that point to determine what would have to be built in order to make that happen. This way of differentiating a business helps Amazon get very clear and vibrant strategic insights for development focused on the needs of their customers: they are customer-centered and have value-driven innovation. The company has demonstrated its commitment to trying new things through initiatives such as Amazon Web Services (AWS) and Amazon Prime. Initially developed as an answer to solve different problems that would arise from the infrastructure that faced Amazon, AWS evolved into a much more sophisticated cloud-computing platform, having experimentally improvised until it reached its final form. In a similar fashion, Amazon Prime started out as a very simple free shipping offer that barely seemed worth it but driven by customer feedback, changing market trends, and the creativeness of Amazon, this evolved as one of their most successful offerings.

4.6.1 The Risks of Over-Planning

Traditional management is often used to create longitudinal, self-explanatory plans, having the premise that future conditions should be more or less stable. But in today's dynamic environment, planning too much can have a detrimental impact on organisational performance. Overemphasising rigid planning can slow an organisation's capability to respond in rapidly unfolding markets and render long-term predictions too quickly outdated due to changing environmental conditions (Grant, 2003; Wolf & Floyd, 2017). Detailed and rigid plans will at one point become obsolete once the unanticipated disturbance or a rapidly changing environment appears, and, in turn, leaving an organisation heavily disadvantaged and slow to react and adapt. Kodak, established in 1888, and Nokia, once leaders in their industry, failed to anticipate and plan for innovations in technology and collapsed. The former having filed for bankruptcy in 2012 after film was replaced by cameras on smartphones. Nokia sold its mobile phone business in 2014, just 22 years after having introduced its first commercially available GSM mobile phone to the market.

Over-planning can affect creativity and innovation since it binds the teams too much into predestined paths that could prohibit them from experimentation and harnessing other potential opportunities. The inflexibility of such a set of strategies will majorly impede the opportunity realisation in fluid market conditions in which flexibility and responsiveness are sensitive and important competitive attributes for an organisation. Moreover, organisations might consume a lot of time and resources in developing exciting plans only to find that the entire plan cannot be implemented, thereby bringing inefficiencies and reducing productivity overall.

Moreover, while over-planning can handicap or disable successful implementation, it can also lead to the analysis paralysis; the postponement of decision-making where people in charge spend their time over-perfecting the best possible plans without taking any action. This may leave a window of opportunity for aggressive competitors to exploit any emerging trends or market dynamics changes. Lastly, while planning is essential to offer structure and direction in businesses, excessive reliance on planning without incorporating an adaptability margin may lead to inabilities to implement plans.

Case Study: Spotify's Agile Working Methods

Spotify is one of the greatest music streaming companies that practice agility across both adaptability and innovation. The Spotify Model, their unique organizational structure, is emphasized by constituent elements (called squad) who are working on different parts of the product they are offering. These squads operate like they were mini startups, offering them the ability to try new things, make decisions, and iterate on their projects. Spotify must have the ability to react immediately to issues relative to user requirements and technology. The organization's business model continuously evolves: Spotify extends its offerings to include podcasts, exclusive content, and AI-driven personalization to stay ahead from competitors. Spotify has shown yet again what resilience is truly about—reinventing along with trends in the market rather than forging its way through resistance. Spotify's agile working model serves as a practical example of shaping agile frameworks to fit a particular company's culture and objectives. By fostering an environment of continuous learning and improvement, Spotify has continued to hold the leadership position in this high competitiveness of the music streaming industry. Their agile way of doing things teaches the importance of forming and allowing a team to decide—empowering them to experiment without fear of having failed to see things work out. This has instilled a culture of research within the company, securing its leader position.

4.6.2 The Value of Experimentation

Experimentation can greatly contribute to the kind of agility that earns strategic success. Organisations can get feedback from the experimentation of new ideas on a smaller scale. In the process, they mitigate the incidence of risks associated with large-scale failures and the utilisation of real-world data for determining business changes.

Eric Ries seminal work (Ries, 2018) has popularised a well-balanced method of planning alongside experimentation development in his Lean Start-up methodology. This innovative approach has set it apart from the traditional planning-motivated strategies. Lean Start-up encourages organisations to create Minimum Viable Products (MVPs) for feedback to their customers or utilise continuous improvements through rapid, constant cycles—all with strong iterations (Fig. 4.4).

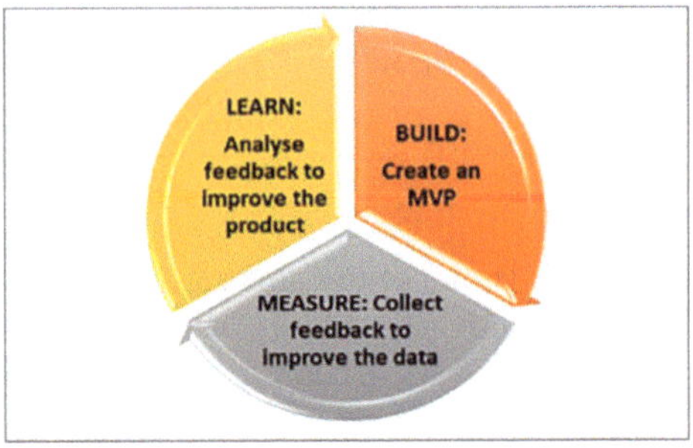

Fig. 4.4 19 Lean Experimentation. Image created by Yioula Melanthiou

4.6.3 Implementing Agile Planning

First steps for agile planning include the adaptation of frameworks such as Scrum or Kanban by organisations. By breaking up one large project into smaller, easily managed tasks, these methodologies assist with the level of agility needed for quick adaptation in a changing environment, resulting also in project management efficiencies.

Minimum Viable Products should be the next step. MVPs help activate the ideas and gather real-world feedback that organisations seek to receive quickly. This also ensures a reduced investment risk, validating the concepts/ideas and modifications based on the feedback of actual users.

Fostering a culture of continuous learning is a must. This develops an environment where creativity thrives, offering employees the space to continuously try new ideas and learn from their failures in the process. This ensures teams have the courage to explore new ways of advancing the organisation's growth while avenues for coping with change and resilience are established.

Alignment of planning cycles to the ground realities of the market is a significant shift. Too-short planning cycles permit such continued and repetitive critical re-examination and fine-tuning to guarantee that plans remain timely in a potential versatile environment. This nature of revisiting plans regularly makes the organisation more aligned with moving markets and better able to leverage new opportunities.

Lastly, organisations need to make measuring outcomes a priority and learning from those outcomes. The use of data and feedback in evaluating the success of an experiment enables the team to make informed decisions and to tweak the strategies they are employing. Since these continuously improve and focus on measurable results, building sustainable growth is achieved by the decision-making process (Fig. 4.5).

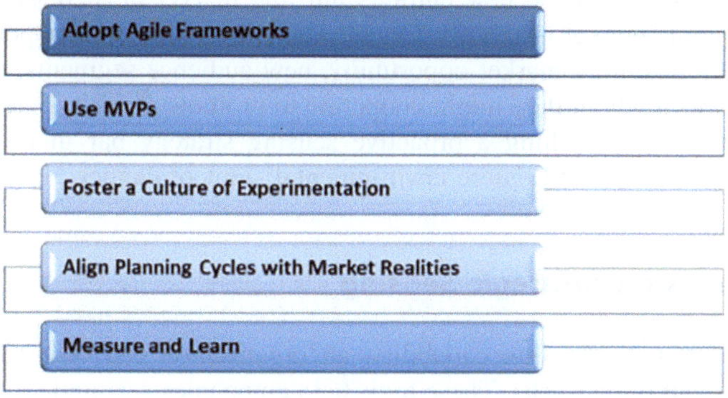

Fig. 4.5 Practical steps to implementing agile planning. Image created by Yioula Melanthiou

4.7 Strategic Sensing: Leveraging Data and Systems for Agility

In the rapid growth and complexity that businesses are experiencing today, strategic sensing has become an essential capability in helping organisations steer through uncertainties, spotting emerging trends, and responding to opportunities and managing disruption before it occurs. Strategic sensing is practiced through vigilant examination of both internal and external factors, identifying threats and opportunities, and shifts in the market environment. Day and Schoemaker (2016) argue that strategic sensing is a most important capability that allows firms to perceive early signals in the environment, be quick in adjusting, and maintain strategic alignment in the face of uncertainty. Organisations are better equipped to know what and how potential changes may occur, and in what direction the implementation of those changes shall be made to enhance resilience and remain competitive.

It is more than what reactive decision-making is. It means being active by gathering data and analysis to make informed business decisions. With the help of the available tools and technologies nowadays, businesses will be able to have real-time information of market changes, enabling them to effectively adapt their strategies accordingly (Chen et al., 2012).

4.7.1 The Importance of Strategic Sensing in Today's Business Landscape

Strategic sensing assists businesses in anticipating upcoming trends, such as changing consumer behaviour, new technologies or innovations, or new legislations affecting businesses (Barrales-Molina et al., 2013). Strategic sensing, furthermore, helps businesses to be informed about potential risks that exist now or in the near future; these risks could be new entrants into the market in increasing competition, or economic market instability. Organisations may also seize opportunities by identifying an emerging market opportunity, new audience segments, or prospective partners with whom they might cooperate to promote growth. Enterprises that do not invest in establishing a proactive sensing strategy bar themselves from competitive advantage and ignore environmental signs of danger.

4.7.2 Tools for Strategic Sensing

To improve strategic sensing capability, an organisation can take advantage of several tools and frameworks. These tools help companies accumulate and analyse different kinds of data regarding the environment they operate in. This data then derives value for decision-making.

Environmental Scanning

Environmental scanning is the systematic monitoring of an organisation's external environment to detect changes in macroeconomics, technology, and socio-political factors that will affect the organisation. This plays a key role in helping companies receive early warnings and getting the strategy adjusted according to external changes; to work closely with these issues, the scan lets enterprises identify and react to weak signals presaging gravitational shifts in the future (Choo, 2001). Thus, it provides an indication of global trends: any such trend is likely to disrupt certain activities of the organisation or lead to a possibly unexpected change in practices that would require them to be conducted differently.

Key areas of monitoring include:

1. Political and Regulatory Changes: Government policies, trade agreements, and regulatory framework changes impact operations of businesses, among others.
2. Economic Trends: Increasing, decreasing, or erratic inflation of prices, changes in interest rates, and fluctuating exchange rates are a few instances that could affect consumer purchases and interruptions in supply chains.
3. Technological Changes: Artificial intelligence, blockchain, and the IoT are among the new technologies that are disrupting sectors while offering new avenues for creativity.
4. Social and Cultural Changes: Changes in consumer value systems, demographics, and lifestyle changes will affect market demand.

By conducting regular environmental scans, organisations can identify potential risks and opportunities early, allowing them to adjust their strategies accordingly.

Competitive Intelligence

Competitive intelligence, sometimes referred to as competition or market intelligence, involves collecting information and making sense of the collected data about competitive environments to forecast shifts within the industry. In a dynamic and competitive world, effective competitive intelligence enables organisations to process market signals at rapid speed and capture strategic insights to enhance their agility and innovation outcomes. This further extends to observing the product launches, marketing strategies created and utilised, as well as partnerships for the purposes of garnering a quantifiable impact from such activities conducted by their rivals, with the ultimate goal of analysing their financial performance through all of these.

Below are the key elements of competitive intelligence:

1. Benchmarking: The comparison of performance by the organisation to that of industry standards and competitors.
2. SWOT Analysis: The identification of the strengths, weaknesses, opportunities, and threats among competitors.
3. Market Positioning: The knowledge of competitors within the market in terms of value propositions and where they place their services to clients.

Competitive intelligence enables any organisation to take critical strategic decisions because it guides the establishment in an in-depth understanding of the competitive landscape. This knowledge enables a business to highlight market gaps and provide a differentiation approach compared to other competitors.

Predictive Analytics

Predictive analytics is the analysis of data, statistical algorithms, and machine learning techniques to detect future possibilities. It may also be described as a prediction of future trends in customer behaviour. Past data may be used as a tool that will allow organisations to recognise such patterns and forecast future outcomes with a high degree of accuracy.

The applications of predictive analytics include:

1. Customer Behaviour Analytics: It pertains to knowing what customers want, their purchase patterns, and how likely they are to switch.
2. Demand forecasting: It includes prediction of future demand of products and services to ensure that optimal inventory and production arrangements are put in place.
3. Risk management: The detection of potential risks and their mitigation strategies before they impact businesses.

With the help of predictive analytics, organisations take proactive actions, optimise operations, and create better experiences for customers. Businesses take full advantage of real-time information to instantly react to changes in market conditions and customer needs.

Case Study: Tesla's Data-Driven Strategy

TESLA

Tesla is an example of how a company applies strategic sensing. The data it collects is the driver behind innovation and customer experience in the company. Performance metrics, usage patterns, and driving conditions are some data points sent to Tesla from the vehicles in real-time. This allows the company to continuously improve its products and services using data.

The company assesses the data from millions of miles driven and refines its algorithms, further improving the safety and performance capabilities of the autonomous driving capabilities of Tesla cars. All this is done while establishing an advantage that keeps the company ahead in the automotive industry and in the race for autonomous vehicles.

Finally, the strategic sensing capability also extends to customer experience. The customers of Tesla are given the opportunity to use its software through over-the-air updates, thus improving their features in real-time while addressing the feedback of customers. This makes it quick for the company to respond to varying customer needs and new market demands.

An organization can improve its agility, resilience against disruption, and competitive advantage in a world that is changing fast by investing in strategic sensing.

4.8 Systems Thinking in Strategy Development

In today's intricate and interconnected world, organisations must realise that their successes hinge upon the ability to prosper within the ecosystem of working, rather than being limited to activities within the organisation. Systems thinking, as a strategic view of the businesses, encourages organisations to view their business environment as an interdependent web of relationships. Adopting a systems-thinking mindset offers companies a better understanding of how varied elements within and outside their organisation interact, influence, and in turn have long-term effects on it.

Systems thinking allows organisations to better understand relations among departments, within a supply chain, and outside stakeholders along with any interdependencies between them (Rădulescu, 2020). Businesses can then improve collaboration through shortening the internal dividers which may separate them.

A systems thinking mindset also accelerates social innovation and adaptability and enhances resiliency by conditioning organisations to pre-empt and mitigate risks before they escalate into serious threats. Businesses are better equipped with effective response strategies to complicated, multi-layered problems, serving to maintain their adaptive competitive edge in a very dynamic business environment.

4.8.1 The Importance of Systems Thinking in Future Organisations

Future organisations functioning on linear, cause-and-effect thought processes amid a changing environment may be limited and unable to progress. Systems thinking provides organisations with the ability to recognise the dynamic interactions that exist in a business ecosystem as well as determine where leverage points exist and where interventions would create positive, systemic transformation.

Take for instance supply chains where disruptions due to geopolitical conflicts, earthquakes, and pandemics can produce cascading disruptions within an organisation. The application of systems thinking can enable a company to notice weaknesses in the supply chain and proceed to necessary changes which will hence augment resilience.

Case Study: Unilever's Holistic Approach to Sustainability

Unilever

Unilever launched its Sustainable Living Plan (USLP) in 2010 with three goals: improvement of health and well-being, reduction of environmental impact, and enhancement of livelihoods inside its value chain. Moreover, the company established itself as a sustainable source by being able to achieve 62% sustainable sources of agricultural materials by the year 2019 and partnered with them for eco-friendly practices. Examples of product innovations include lower-temperature wash detergents to save energy, installations for reductions in plastic use, and increases in recyclability in packaging.

They also addressed social impact for instance through Lifebuoy's global hand washing campaigns that were great at preventing diseases, or Dove's Self-Esteem Project aimed at empowering millions of young people. Programs such as this one will help gender inequality issues and focus on encouraging women's entrepreneurs improve their livelihoods. By 2020, this Sustainable Living Plan had made it possible for 1.3 billion people to lead a better life in terms of health, reduced their product carbon footprint by one-third, and positively impacted 2.34 million lives in its supply chain. Sustainable brands grew 69% faster than all other brands and accounted for 75% of overall growth.

4.8.2 Principles of Systems Thinking

One of the most important features of systems thinking is the principle of interdependence. Interdependence is a recognition of the fact that all parts of a system are connected. Any alteration made in one part of a system will invariably have repercussions on the other parts. Hence, the relationships need to be well understood because they determine how the overall behaviour can be influenced.

On the bases of this, a holistic view is also necessary to analyse how the whole system creates impact with its interaction to the external environment. Organisations can unveil the much deeper effect through which a few components combine, and where to improve performance. The essential tool within this approach is feedback, describing how activity in the system disturbs the current equilibrium or builds up over time or drives activity forward or backward. By recognising and manipulating these loops, the system remains sensitive and level. In addition to this, a good system must also accommodate this kind of adaptability to continuous learning and change because flexibility and willingness to change are the keys to growth in environments where changes take place and to keeping such systems alive for the long run. These principles together make up a unified understanding framework for complex systems.

4.8.3 Implementing Systems Thinking

The very first step in systems thinking implementation is first mapping the organisational ecosystem. That is, identifying key stakeholders, processes, and interdependencies that define how these different elements interact with one another and influence one another. Clear mapping of these relationships unveils the complexity of the system and espouses areas for further improvement. Another fundamental step is the recognition of feedback loops. Feedback loops can either reinforce changes made, thus causing them to grow larger, or counter them, thus creating system stability. Understanding the loops helps an organisation to know better how to make predictions about outcomes, and it can subsequently adjust its strategies appropriately.

Cross-functional collaboration is essential in addressing the kinds of complex challenges the system presents. By stimulating cross-functional cooperation, organisations can encourage broader-perspective thinking and build richer solutions. Data and analytics are necessary in systems thinking. Organisations harness data so that they can look at changes, establish patterns, and conclude data-driven, system-aligned decisions. Data accuracy guarantees an even harder insight into environmental dynamics and, moreover, prepares the ground for proactive management.

Finally, developing a learning culture to build adaptability in the organisation is of paramount importance. It helps keep teams flexible and learning all the time, which prepares them to respond most effectively to changes. With a systems thinking approach, organisations can enhance their strategic agility, increase resilience, and allow for sustainable growth in a rapidly changing business landscape. The success of Unilever illustrates the value of integrating systems thinking within core business strategies to create awareness and address complex and interconnected challenges.

4.9 Corporate Renaissance: Balancing Tradition and Innovation

Corporate Renaissance signifies the renewed emphasis on creativity, innovation, and continuous reinvention in the business world today. In such a future world, creativity becomes more a matter of necessity than a luxury for survival and growth. For any organisation, the high risk of becoming obsolete exists if it ceases to innovate, but a successful organisation can combine tradition with a forward-thinking approach and adapt to new market demands to secure the organization in its long-term success.

Changing technology, changing tastes, and increasing competition are the forces of this Corporate Renaissance. New businesses will need culture that supports radical thinking while maintaining some of their core values as requirements of relaying relevancy in a fast-paced world.

4.9.1 The Balance of Tradition and Innovation

Strategic agility presents another dimension: balancing tradition and innovation. Old and traditional organisations with flagship brands must learn to evolve but not lose their core values. Innovation must not only be the forte of the company; it must also extend to continuity and stability to secure the identity and the credibility with customers. That balance must ensure that organisations are prepared to embrace changes in the market and still retain the trust and loyalty of their stakeholders (Gil et al., 2024; Sundaramurthy & Kreiner, 2008).

Many renowned organisations have achieved this course by embracing innovation in culture, yet it is consistent with their fundamental underpinnings. The case study of Google will demonstrate how two of the world's most iconic companies have managed to stay relevant through balancing their proverbial feet in tradition and innovation.

Case Study: Google's Innovation Culture

One company that clearly illustrates the importance of fostering a culture of innovation to remain competitive is Google. Even from its inception, Google's key drivers of success were creativity and experimentation. One of its well know initiatives is the '20% time' program which allows its employees to use a portion of their work week in projects they are passionate about as it believes that passion can often result in great products. It further allows them to experiment, take risks, and learn from failures quickly. Indeed, this program combining innovation and creativity, resulted in the development of several successful products like Gmail, Google Maps and Google News. Their innovation culture has also radiated beyond their well-known core business, that of search alone. Through the umbrella company, Alphabet, the company ventured into other types of businesses such as cloud computing, artificial intelligence, and autonomous vehicles. Each of these efforts clearly shows Google's effort to always remain at the forefront or present in emerging technological landscape trends and different avenues for business.

4.9.2 The Broader Impact of the Corporate Renaissance

The Corporate Renaissance has a much broader effect that extends well beyond technology companies; businesses in healthcare, finance, retail, manufacturing, and in other sectors are carefully infusing aspects of innovation and creativity, reshaping their historically established business models. This cross-sectoral transformation aligns with research on industry-wide digital innovation, which shows that traditional firms are increasingly investing in digital infrastructure, ecosystem partnerships, and data-driven strategies to enhance resilience and responsiveness to disruption (Yoo et al., 2012). Adopting these changes and shifting towards actively supporting continuous innovation are seen as robust predictors of survival. This supports the idea that sustained innovation is a dynamic capability necessary for the survival of any company struggling to maintain its position in turbulent environments (Teece, 2007). Improving their digital footprint, for example, has proven to be a successful transition for the traditional retailer Carrefour, owing to the significant investment in e-commerce solutions in addition to the powerful data analytics that enhances their customer experience. In the same way, healthcare organisations are adopting even more digital technologies to enhance patient care and operations and ensure better healthcare for their patients.

This transformation has introduced possibilities for completely different operational processes but has also transformed the mindset of operations. Firms holding

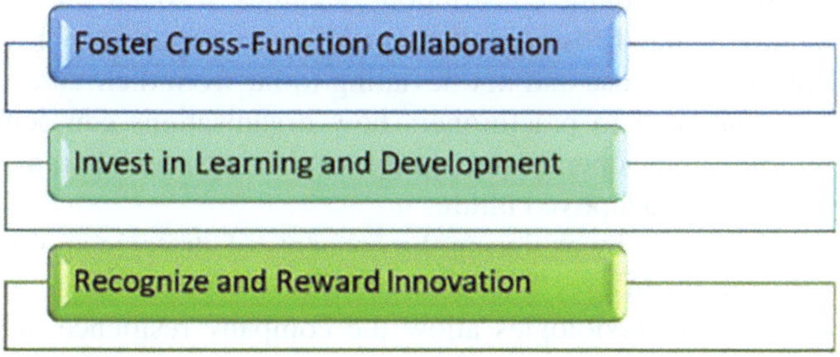

Fig. 4.6 Practical steps to build a culture of innovation. Image created by Yioula Melanthiou

on to culture and enabling creativity with flexibility and sustained improvements will be more capable of embracing the complexities today and into the future while creating sustainable growth paths. This Corporate Renaissance compels businesses to develop beyond the old practices and root innovation within their critical strategies for competition.

The building of this innovative culture now requires cross-functional teams to collaborate in creating and developing innovations. According to research, giving teams authority and encouraging experimentation makes them feel ownership about the results of the experiments, and this results in creating much better innovative performance (Birkinshaw et al., 2008). This, in turn, emphasises the role of learning and development, as empowering employees with the right tools and resources can foster the development of such innovations. Awarding reward and recognition for innovation also nurtures an ideal environment for continuous improvement. It creates an atmosphere ripe for flourishing innovation by introducing success and learning from failure, to build a resilient and future-ready business.

Finding the right balance between traditional and forward-thinking approaches will allow business to position themselves effectively into this new era. The right balance will ensure that businesses are resilient and agile but are also true to their traditions and core values. Those businesses that will be able to successful adapt well during this dynamic period, will be best prepared to capitalise on new opportunities and mitigate foreseen challenges with the maximum success and grace, and will be able to build innovations—if they embed those innovations into their core operations themselves—in the form of reward engaging in some new opportunity while mitigating the impacts of newly emerging challenges (Fig. 4.6).

4.10 Conclusion: Dilemmas, Disruptions, and the Future of Strategic Agility

In a world characterised by endless change, complexity, and uncertainty, strategic agility has increasingly become critical for organisations. The future organisation encounters increasing disruptions from the evolving technological landscape, geopolitical conflicts, economic constrictions, changes in social values, and so on. These increasingly demand fostering the capability to respond quickly while preserving their primary purpose and not deviating to far from their core values. This chapter has demonstrated to practitioners how organisations can achieve strategic agility by embedding operational agility, exploiting critical frameworks, and nurturing an innovation-adaptive culture.

The chapter also began by exploring the concepts of strategic agility and adaptability using Toyota as an excellent illustration of how continuous improvement and lean production methodologies allow the company resilience in the face of increased disruptions. The success of Toyota reveals that agility is not reactive but requires proactive adjustments to processes and strategies. Companies should build flexibility in their operations, anticipating market movements, decrease work inefficiencies, and remain competitive in changing market environments.

Organisations must scan their external environment for political, economic, social, technological, legal, and environmental elements. Geopolitical conflicts, like the war between Russia and Ukraine, have forced companies to re-evaluate their sourcing strategies and diversify supply chains to better mitigate risks. Another example is Netflix, which shifted from its original model of DVD rentals to streaming services, an indication of perceptual changes in technology and society that businesses could expect to stay ahead of the competition. The ability to sense and respond to external changes is a critical feature of agility that gives businesses the potential to seize an opportunity and shield itself from possible threats.

It elaborated further in the chapter that resilience, agility, and sustainability would be the three attributes that determine long-term success. Companies will not only have to respond to disruptions but also create systems that absorb shocks and bounce back faster. IKEA demonstrates how sustainability efforts, by transforming a company into a circular economy integrated with investments in renewable energy, can make companies more resilient as well as responsive to changing societal values. IKEA has successfully integrated sustainability into its core activities, and this shows how companies can achieve agility throughout all challenges facing society and the environment.

Corporate Renaissance is about balancing tradition and innovation within organisations so that they remain relevant in fast and ever-changing markets. Companies like Apple and Google are an example of such a balancing act in that they remain faithful to their core value systems, while encouraging an innovative culture within the company. Apple has always been design simple, user-experience oriented, in launching disruptive inventions like the iPhone. Likewise, Google's innovation culture, driven by their policy of "20% Time" and a "Fail Fast" approach, has yielded the birth of such revolutionary products as Gmail and Google Maps. Agility requires that companies evolve yet stay true to core values.

The chapter also focused on how purpose-driven organisations can achieve agility. A purpose-driven company can align its strategy with stakeholder expectations and adapt to change without compromising its core values. Alibaba's Rural Taobao initiative illustrates digital inclusion for social and business value, showing how companies that make purpose a priority more often races steadfast towards sustainable growth. It gives guiding frameworks for making decisions, which ultimately will help in remaining agile when things change in the markets.

Authentication and trust are imperative for driving agility. Patagonia, which donates profits for environmental advocacy, has created loyal clientele and built its own resilience. Trust becomes important in stormy markets as a strategical asset with which companies can achieve stakeholder loyalty, reduce their exposure, and speed up decision-making in uncertain times.

This chapter further highlighted the importance of balancing planning with experimentation to achieve agility. Conventional long-range planning interferes with adaptability, while too much short-termism risks orientation loss. Models used by Spotify and Amazon show the benefits of agile planning by allowing teams to experiment, iterate, and innovate. Their agile planning strategies and actions prioritise continuous improvement while being customer-centric, creating a balance between a long-term vision with short-term flexibility.

Finally, strategic sensing and systems thinking are critical capabilities for enhancing agility within organisations. Businesses should gain deep understanding of both external and internal environments to continuously identify emerging trends and risks. One good example of strategic sensing would be Tesla's use of data to enhance autonomous driving and customer experience. By tracking data in real time, Tesla refines its offerings and makes sure that it stays ahead of rivals. Similarly, Unilever's value systems thinking includes the integration of social, environmental, and economic dimensions in addressing complex societal issues. Firms can invent solutions and build resilience against future disruptions when they understand their interdependencies in ecosystems.

Concluding, strategic agility should be regarded by firms as a core competence in managing their businesses. This calls for a shift from rigid planning to continuous learning, experimentation, and collaboration. Organisations that foster value-based culture, trust, and systems thinking can build resilience, improve adaptability, and create enduring and sustainable value for stakeholders. Strategic agility becomes a state of being rather than a management tool in helping companies thrive amid a milieu of uncertainty and complexity.

Future Management in Practice: Strategic Agility

This chapter's focus has been on enhancing *Organisational Capability.* Future Management strategists do not regard agility as a fad, but as an essential skill for companies faced with continuous difficulty, or a permacrisis. Success, from the lens of traditional management, has stability, planning, and control. Future Management acknowledges a different reality: to survive and flourish, companies need to sense and adapt to changes and refine their evolution.

1. Agility as a Foundational Skill

Leaders must integrate agility into their organisations from the outset. This also involves crafting organisational cultures and structures that facilitate swift learning, design, and decision-making at the edge, and adaptive resource reallocation. Agility must be managed as a key asset, not a casual benefit.

2. Integrating Planning with Experimentation

Future Management recognises that over-planning creates rigidity, while under-planning creates chaos. The managerial task is to strike a balance by setting direction with clarity of purpose while leaving room for experimentation, iteration, and course correction. This demands leaders who are comfortable with ambiguity and capable of making timely decisions without complete certainty.

3. Relating Agility to Resilience and Sustainability

Agility is not about reacting quicker just to react. It produces results only when it is paired with resilience and sustainability. Resilience, as defined in Chap. 1, gives the capability to endure shocks, while sustainability guarantees that the responses taken enhance long-term viability. There must be an evolution from and beyond quick performance to agile actions, to take measures for organisational futures that are stronger and permanent.

4. Strategic Sensing in Action

The strategic sensing is crucial in permacrisis, as we have seen in Chap. 1. In this chapter, we learn how Future Management employs this methodology: by integrating sensing into day-to-day activities to help organisations detect weak signals promptly, respond to them swiftly, and ensure these responses are in tune with the organisation's enduring strategy.

5. Purpose As a Guide

The leadership purpose defined in Chap. 2 becomes the compass for agility. Agility lacking that anchor runs the risk of degenerating to opportunism with fast and shallow reactions. Future Management highlights the need for a purpose to actively steer swift choices, so that shifts are purposeful and in tune with values and enduring prosperity.

In Practice:
Future Management is to nurture agility as an organisation-wide proficiency rooted in a purpose, guided by principles, and exercised with resilience. Managers who accept agility in the described manner will, not only react to challenges, but

will use them in propelling the organisation towards innovation and competitive leverage that is sustainable.

MANAGERS MUST BALANCE PLANNING AND EXPERIMENTATION, BUILD STRATEGIC SENSING, AND ANCHOR AGILITY IN PURPOSE.

References

Argyris, Chris and Schön, D.A., (1997) Organizational learning: A theory of action perspective. *Reis*, (77/78), pp.345 348.

Avolio, B. J., & Gardner, W. L. (2005). Authentic leadership development: Getting to the root of positive forms of leadership. *The Leadership Quarterly, 16*(3), 315–338.

Bachmann, R., & Inkpen, A. C. (2011). Understanding institutional-based trust building processes in inter-organizational relationships. *Organization Studies, 32*(2), 281–301.

Barney, J. B. (2021). *Gaining and sustaining competitive advantage* (5th ed.). Pearson.

Barrales-Molina, V., Bustinza, Ó. F., & Gutiérrez-Gutiérrez, L. J. (2013). Explaining the causes and effects of dynamic capabilities generation: A multiple-indicator multiple-cause modelling approach. *British Journal of Management, 24*(4), 571–591.

Bartlett, C. A., & Ghoshal, S. (1995). Changing the role of top management: Beyond structure to processes. *Harvard Business Review, 73*(1), 86–96.

Birkinshaw, J., Hamel, G., & Mol, M. J. (2008). Management Innovation. *Academy of Management Review, 33*(4), 825–845.

Boin, A., Overdijk, W., Kuipers, S., & van den Oord, A. (2021). Resilience and the future of crisis management. *Journal of Contingencies and Crisis Management, 29*(1), 3–10. https://doi.org/10.1111/1468-5973.12347

Bronlet, X. (2023). Business agility: Does the size really matter? *International Journal of Economics, Business and Management Research, 07*(03), 97–112.

Bughin, J., Catlin, T., LaBerge, L., & Mellbye, A. (2021). *The state of AI-driven digital transformation.* McKinsey & Company. https://www.mckinsey.com

Cascio, Jamais, Johansen, Bob and Williams, Angela F., 2025. *Navigating the Age of Chaos: A Sense-Making Guide to a BANI World That Doesn't Make Sense.* Berrett-Koehler Publishers.

Chen, H., Chiang, R. H., & Storey, V. C. (2012). *Business intelligence and analytics: From big data to big impact. MIS quarterly* (pp. 1165–1188).

Choo, C. W. (2002). Environmental scanning as information seeking and organizational knowing. University of Amsterdam, Netherlands. *Sprouts: Working Papers on Information Systems, 2*(1), 2–1.

Christopher, M., & Peck, H. (2004). Building the resilient supply chain. *The International Journal of Logistics Management, 15*(2), 1–14. https://doi.org/10.1108/09574090410700275

Coutu, D. L. (2002). How resilience works. *Harvard Business Review, 80*(5), 46–55.

Day, G. S., & Schoemaker, P. J. (2016). Adapting to fast-changing markets and technologies. *California Management Review, 58*(4), 59–77.

Dirks, K. T., & Ferrin, D. L. (2001). The role of trust in organizational settings. *Organization Science, 12*(4), 450–467.

Duchek, S. (2020). Organizational resilience: A capability-based conceptualization. *Business Research, 13*, 215–246. https://doi.org/10.1007/s40685-019-0085-7

Euromonitor International. (2021). *Plant-based eating trends: Market growth and consumer preferences.* https://www.euromonitor.com

European Commission. (2023). *Ukraine's trade and economic challenges amid conflict.* European Union Publications Office. https://ec.europa.eu

European Data Protection Board. (2023). *Annual report on GDPR enforcement and compliance.* https://edpb.europa.eu

Gardner, W. L., Avolio, B. J., Luthans, F., May, D. R., & Walumbwa, F. (2005). "Can you see the real me?" A self-based model of authentic leader and follower development. *The Leadership Quarterly, 16*(3), 343–372.

Gil, M., Thor, K., & Gemheden, A. (2024). Managing the tensions between tradition and innovation in family firms. *Journal of Family Business Management, 15*(5)

Girotra, K., & Netessine, S. (2014). Four paths to business model innovation. *Harvard Business Review, 92*(7). https://hbr.org/

Grant, R. M. (2003). Strategic planning in a turbulent environment: Evidence from the oil majors. *Strategic Management Journal, 24*(6), 491–517.

Hamel, G., & Valikangas, L. (2003). The quest for resilience. *Harvard Business Review, 81*(9), 52–63.

Henderson, R., & Steen, E. V. D. (2015). Why do firms have "purpose"? The firm's role as a carrier of identity and reputation. *American Economic Review, 105*(5), 326–330.

Ilies, R., Morgeson, F. P., & Nahrgang, J. D. (2005). Authentic leadership and eudaemonic well-being: Understanding leader–follower outcomes. *The Leadership Quarterly, 16*(3), 373–394.

International Monetary Fund. (2023). *Global economic impact of the Russia-Ukraine war.* https://www.imf.org

Kernis, M. H., & Goldman, B. M. (2006). A multicomponent conceptualization of authenticity: Theory and research. *Advances in Experimental Social Psychology, 38*, 283–357.

Lengnick-Hall, C. A., Beck, T. E., & Lengnick-Hall, M. L. (2011). Developing a capacity for organizational resilience through strategic human resource management. *Human Resource Management Review, 21*(3), 243–255. https://doi.org/10.1016/j.hrmr.2010.07.001

McKinsey & Company. (2021). *The future of digital resilience in business operations.* https://www.mckinsey.com

McKinsey & Company. (2022). *The future of mobility and the gig economy.* https://www.mckinsey.com

McKinsey & Company. (2023). *Managing supply chain risks in a geopolitically unstable world.* McKinsey Global Institute. https://www.mckinsey.com

McKinsey Global Institute. (2022). *The state of AI and predictive risk management.* https://www.mckinsey.com

Mintel. (2022). *Global consumer trends report: The rise of ethical and sustainable consumption.* https://www.mintel.com

O'Reilly, C. A., III., & Tushman, M. L. (2013). Organizational ambidexterity: Past, present, and future. *Academy of Management Perspectives, 27*(4), 324–338.

Radović-Marković, M., Marković, D., Vučeković, M., & Chou, Z. (2023). Strategic organizational resilience during and after the covid-19: Case study of western Balkans. *International Review, 3–4*, 15–24.

Rădulescu, C. M., Slava, S., Rădulescu, A. T., Toader, R., Toader, D. C., & Boca, G. D. (2020). A pattern of collaborative networking for enhancing sustainability of smart cities. *Sustainability, 12*(3), 1042.

Ramírez, Rafael, and Wilkinson, Angela (2016). *Strategic reframing: The Oxford scenario planning approach.* Oxford University Press.

Reeves, M., & Whitaker, K. (2020). A guide to building a more resilient business. *Harvard Business Review, 2.* https://hbr.org/

Ries, E. (2016). *The lean Start-up by Eric Ries.* Crown Business.

Ries, Eric, 2018. *The lean startup*(Republish). Bentang Pustaka.

Sancak, I. E. (2023). Change management in sustainability transformation: A model for business organizations. *Journal of Environmental Management, 330*, Article 117165.

Sheffi, Y. (2015). *The power of resilience: How the best companies manage the unexpected*. MIT Press.

Stacey, Ralph D. (1995) The science of complexity: An alternative perspective for strategic change processes. *Strategic Management Journal 16*(6): 477-495.

Stoll-Kleemann, S., & Schmidt, U. J. (2022). Transforming food systems: Consumer adoption of plant-based diets. *Environmental Research Letters, 17*(3), Article 034028.

Snowden, D. J., & Boone, M. E. (2007). A leader's framework for decision making. *Harvard Business Review, 85*(11), 68.

Sundaramurthy, C., & Kreiner, G. E. (2008). Governing by managing identity boundaries: The case of family businesses. *Entrepreneurship Theory and Practice, 32*(3), 415–436.

Sutcliffe, K. M., & Vogus, T. J. (2003). Organizing for resilience. In K. Cameron, J. E. Dutton, & R. E. Quinn (Eds.), *Positive organizational scholarship* (pp. 94–110). Berrett-Koehler.

Tallon, P. P. (2021). Digital resilience: Why IT and data are more critical than ever. *MIS Quarterly Executive, 20*(2), 147–159. https://doi.org/10.17705/2msqe.00046

Teece, D. J. (2007). Explicating dynamic capabilities: The nature and microfoundations of (sustainable) enterprise performance. *Strategic Management Journal, 28*(13), 1319–1350.

UK Parliament. (2022). *Brexit and the UK construction industry: Labour shortages and policy adjustments*. House of Commons Business Report. https://www.parliament.uk

Van der Vegt, G. S., Essens, P., Wahlström, M., & George, G. (2015). Managing risk and resilience in supply chains. *Academy of Management Journal, 58*(4), 971–980.

Williams, A., Whiteman, G., & Kennedy, S. (2021). Cross-scale systemic resilience: Implications for organization studies. *Business & Society, 60*(1), 95–124.

Wolf, C., & Floyd, S. W. (2017). Strategic planning research: Toward a theory-driven agenda. *Journal of Management, 43*(6), 1754–1788.

World Bank. (2022). *Trade disruptions and economic fallout from the Russia-Ukraine conflict*. World Bank Policy Research. https://www.worldbank.org

World Economic Forum. (2022). *Global business regulations: Adapting to legal and policy shifts in a digital world*. https://www.weforum.org

World Trade Organization. (2023). *Impact of sanctions on global trade and economic stability*. WTO Trade Reports. https://www.wto.org

Yoo, Y., Boland, R. J., Jr., Lyytinen, K., & Majchrzak, A. (2012). Organizing for innovation in the digitized world. *Organization Science, 23*(5), 1398–1408.

MIX
Papier aus verantwortungsvollen Quellen
Paper from responsible sources
FSC® C105338

If you have any concerns about our products,
you can contact us on
ProductSafety@springernature.com

In case Publisher is established outside the EU,
the EU authorized representative is:
Springer Nature Customer Service Center GmbH
Europaplatz 3, 69115 Heidelberg, Germany

Printed by Libri Plureos GmbH
in Hamburg, Germany